T0212075

Introduction to Number Theory

Introduction to Number Theory covers the essential content of an introductory number theory course including divisibility and prime factorization, congruences and quadratic reciprocity. The instructor may also choose from a collection of additional topics.

Aligning with the trend toward smaller, essential texts in mathematics, the author strives for clarity of exposition. Proof techniques and proofs are presented slowly and clearly.

The book employs a versatile approach to the use of algebraic ideas. Instructors who wish to put this material into a broader context may do so, though the author introduces these concepts in a non-essential way.

A final chapter discusses algebraic systems (like the Gaussian integers) presuming no previous exposure to abstract algebra. Studying general systems helps students to realize unique factorization into primes is a more subtle idea than may at first appear; students will find this chapter interesting, fun and quite accessible.

Applications of number theory include several sections on cryptography and other applications to further interest instructors and students alike.

Textbooks in Mathematics

Series editors:
Al Boggess, Kenneth H. Rosen

https://www.routledge.com/Textbooks-in-Mathematics/book-series/CANDHTEXBOOMTH

Introduction to Number Theory

Mark Hunacek

CRC Press
Taylor & Francis Group
Boca Raton London New York

CRC Press is an imprint of the
Taylor & Francis Group, an **informa** business

A CHAPMAN & HALL BOOK

First edition published 2023
by CRC Press
6000 Broken Sound Parkway NW, Suite 300, Boca Raton, FL 33487-2742

and by CRC Press
4 Park Square, Milton Park, Abingdon, Oxon, OX14 4RN

CRC Press is an imprint of Taylor & Francis Group, LLC

© 2023 Mark Hunacek

Reasonable efforts have been made to publish reliable data and information, but the author and publisher cannot assume responsibility for the validity of all materials or the consequences of their use. The authors and publishers have attempted to trace the copyright holders of all material reproduced in this publication and apologize to copyright holders if permission to publish in this form has not been obtained. If any copyright material has not been acknowledged please write and let us know so we may rectify in any future reprint.

Except as permitted under U.S. Copyright Law, no part of this book may be reprinted, reproduced, transmitted, or utilized in any form by any electronic, mechanical, or other means, now known or hereafter invented, including photocopying, microfilming, and recording, or in any information storage or retrieval system, without written permission from the publishers.

For permission to photocopy or use material electronically from this work, access www.copyright.com or contact the Copyright Clearance Center, Inc. (CCC), 222 Rosewood Drive, Danvers, MA 01923, 978-750-8400. For works that are not available on CCC please contact mpkbookspermissions@tandf.co.uk

Trademark notice: Product or corporate names may be trademarks or registered trademarks and are used only for identification and explanation without intent to infringe.

ISBN: 9781032332055 (hbk)
ISBN: 9781032017204 (pbk)
ISBN: 9781003318712 (ebk)

DOI: 10.1201/9781003318712

Typeset in Palatino
by codeMantra

This book is dedicated to Leslie, Adrienne and Sofia,

the three most important women in my life.

Contents

Preface

This book, intended as a text for a junior/senior-level undergraduate course in elementary number theory, is based on my experience teaching such a course at Iowa State University. The course, though taught by a member of the mathematics department, is cross-listed with the computer science department so the audience typically consists of mathematics and computer science majors, in roughly equal proportion, along with an occasional minor in one of these subjects.

Both the computer science and mathematics departments offer an "introduction to proofs" course, completion of either one of which is the only prerequisite for the number theory course. Despite this requirement, however, I have found over time that any real level of comfort with creating proofs cannot be assumed, so I have spent at least one class period reviewing this material. This review is reflected in this text: there is an Appendix on proof techniques, and particularly in the beginning of the text, proofs are presented in considerable detail.

Another issue that I grappled with when teaching the course is the extent of algebra that I wished to include. Abstract algebra not being a prerequisite for the course, most students in it had never heard of words like "group", "ring" or "field". The first time I taught the course, I likewise avoided any mention of these terms, but I found that maddingly frustrating. I was reminded of my experience as an undergraduate taking a comparable course; when the subject of primitive roots came up, I was able to understand the definition and the various proofs, but I had little intuitive feel for the idea; it wasn't until later, after learning what a group was, that I realized that all of this was just about cyclic groups. Likewise, results like Euler's theorem suddenly became much clearer to me, when I realized the "right" context for these results. So, teaching the course in subsequent semesters, I experimented with mentioning enough algebra to at least give the students some indication of the fact that these results were best viewed in a more general context. Some semesters I would just mention the technical terms and tell the students without detail that there was something deeper going on; on other occasions, I would take a day or two to actually develop some abstract algebra in class and then show, for example, the connection between Euler's theorem and Lagrange's theorem in group theory. All these approaches have their benefits and drawbacks, and to accommodate differing choices among instructors, I have tried to provide flexibility in this text. The book can be read without ever mentioning abstract groups, rings or fields, but these terms are introduced in an Appendix and at least referred to (in a non-essential way) in the text. An instructor can simply ignore these references, or discuss them in varying degrees of detail, as he or she sees fit.

One inclusion of an algebraic idea that I could not resist occurs in the section on greatest common divisors. I have always had a fondness for proving the existence of the gcd by using ideals, so, in Chapter 1, I define that concept (for the integers only), prove that any ideal in the integers is generated by a single element, and use that result to quickly prove, in one fell swoop, that the gcd of two integers exists and is a linear combination of those integers.

This approach to the gcd pays dividends in the final chapter of the book, which also introduces some algebraic ideas, though in a fairly concrete setting, focusing on specific examples rather than abstract algebraic systems. This chapter begins with a fairly detailed look at the Gaussian integers, mimicking, wherever possible, the various arguments used previously in the text for the ordinary integers (including the concept of an ideal and using ideals to prove the existence of a gcd). From the Gaussian integers, we proceed to other quadratic extensions, including a discussion of algebraic systems in which unique factorization fails, thus showing the students that unique factorization is a more subtle concept than might have originally been thought.

As a very pleasant additional benefit, studying other algebraic systems can actually be used to prove results *about* the ordinary integers. As seen in the text, for example, the Gaussian integers can actually be used to prove results about sums of two squares of integers and also used to classify Pythagorean triples. Studying the quaternions allows a proof that any positive integer can be written as the sum of four squares. Over the years, I have found that my students find this material to be interesting, fun and quite accessible. And here again, the instructor has some discretion in determining whether to use algebraic terms like "ring" and "field"; I have written the book so as to accommodate either choice.

In writing this book, I have resisted the urge to discuss a plethora of topics, most of which will never be gotten to in a one-semester introductory course. I find it discouraging to use a book as a text for a course and then only cover half (or less) of it. Students, I think, don't like this either, particularly since they are the ones who are paying for the book. Therefore, I have tried to write a book that covers the essential content of an introductory number theory course (divisibility and prime factorization, congruences, quadratic reciprocity) and a collection of topics from which the professor can choose (perfect numbers, sums of squares, Pythagorean triples, primitive roots and, as previously noted, a chapter on algebraic systems other than the integers). Because I invariably had computer science majors in my class, and because the math majors also generally found it interesting, I also have included some optional material on cryptography. All told, there is probably a little more material in the text than can be covered in a one-semester course, but not so much more as to be discouraging. The last time I taught the course, I covered in one class period a selection of material from Chapter 0, and then did Chapters 1 through 6 in their entirety (weaving in, as appropriate, the three Appendices). This left me enough time to cover a substantial amount of

Chapter 7. I have never succeeded in covering the quaternions, but I always make it a point to get to Section 7.9 and at least give an example or two of non-unique factorization in a quadratic extension of the integers. Because the course more or less begins with unique factorization in the integers, coming full circle and looking at non-unique factorization in other contexts seem an excellent way to end the semester.

Author

Mark Hunacek has advanced degrees in both mathematics (PhD, Rutgers University) and law (JD, Drake University Law School). He is now a Teaching Professor Emeritus at Iowa State University, and before entering academia, he was an Assistant Attorney General for the state of Iowa.

0

Introduction: What Is Number Theory?

Carl Friedrich Gauss, whom many people consider the greatest mathematician who ever lived, once described number theory as the "Queen of Mathematics". Indeed, the integers ("whole numbers"), and the patterns they exhibit, have been the subject of fascination and study for literally thousands of years. Euclid's famous treatise *The Elements*, which is often thought as being solely related to geometry, actually contains many results that are theorems of number theory. In Chapter 1 of this text, for example, we will give Euclid's proof that there are infinitely many prime integers.

For our purposes, the term "number theory" will (mostly) refer to the study of the integers and various issues connected with them. Unlike many areas of mathematics, where the problems and conjectures themselves (let alone the proofs) are so technical that one has to be a specialist in the area to even understand them, many questions in number theory do not involve technical terms or results and can be understood by a grade-schooler. In this introductory chapter, we will look at some examples of these problems, so as to try and give a sense of the flavor of the subject.

First, let us start with one of the most famous problems in mathematics: *Fermat's Last Theorem*. This is one of the great success stories of mathematics, and it also has a fascinating history that dates back to the Pythagorean theorem: if a right triangle has sides of length x and y and hypotenuse of length z, then $x^2+y^2=z^2$. From a number-theoretic point of view, it is of interest to look for positive *integer* solutions to this equation, such as $x=3, y=4$ and $z=5$. Later in the text, we will find the general form of *all* such solutions.

Mathematicians love to generalize from one problem to another, and when you've considered $x^2+y^2=z^2$, it is not too big a leap to consider more general equations like $x^n+y^n=z^n$, where $n>2$ is a positive integer. Equations like these are called *Diophantine* equations, in honor of the Greek mathematician Diophantus, who wrote a textbook titled *Arithmetica* in which he discussed solutions of many equations.

In the mid-1600s, a lawyer and amateur mathematician named Pierre Fermat was reading Diophantus's book and wrote in the margin that he had discovered a "marvelous proof" that "this margin is too narrow to contain" that for positive integers $n>2$, the above-mentioned equation $x^n+y^n=z^n$ has no solution in positive integers x, y and z.

Whether Fermat actually did have such a proof is something that will never be known for sure, but most authorities believe that he did not. In any event, this cryptic marginal reference led to a search, lasting for more than

DOI: 10.1201/9781003318712-1

300 years, for a proof of this result. Correct proofs were given for some specific values of n (Fermat himself proved the result for $n=4$, and Euler proved the result for $n=3$ by a different method) but nobody came up with a proof that worked for all n. On the other hand, nobody could come up with a counterexample showing the result to be false. Some mathematicians, including some very good ones, thought they had come up with a proof, but errors were always found. Sometimes these errors themselves shed some light on subtle points, such as the uniqueness of factorization into primes.

Finally, in 1993, Andrew Wiles announced that, after 7 years of intense effort, he had found a proof of the result. Unfortunately, Wiles' proof was found to contain an error, but that error was, in collaboration with his student Richard Taylor, eventually patched up in 1994. The proof, published in 1995, uses very deep and difficult mathematics that is far beyond the scope of this text, and which did not even exist in Fermat's time.

As just noted, Fermat's equation $x^n+y^n-z^n=0$ is an example of a Diophantine equation. More generally, a Diophantine equation in k variables $x_1, x_2, ..., x_k$ is an equation of the form $p(x_1, x_2, ... x_k)=0$, where p is a polynomial in these variables with integer coefficients. The study of such polynomial equations, which is a major part of number theory, itself naturally leads to several questions, including questions of interest to computer scientists. Specifically, one might ask: Is there an algorithm for determining whether a given Diophantine equation has a solution in integers or rational numbers? If so, is there an algorithm for determining all such solutions? If we can't determine *all* solutions, can we at least determine *some*? The first of these questions is known as Hilbert's Tenth Problem, so named because it was the tenth of 23 then-open problems identified by the mathematician David Hilbert in a famous speech that he gave in Paris in the year 1900. In 1970 it was shown, via the collaborative efforts of several mathematicians, that no such algorithm exists. This result shows just one way in which number theory intersects with other areas of mathematics (here, logic).

Fermat's Last Theorem and Hilbert's Tenth Problem are at least mathematical problems that were eventually solved. There are other problems in number theory that remain unsolved to this date. A number of them are also easy to state. As a first example, let us say that a positive integer n is *perfect* if the sum of its factors (other than n itself) is equal to n. For example, 6 is perfect, because $1+2+3=6$. There are, as of this writing, only 51 known perfect numbers, all of which are even. This suggests two questions: Are there any odd perfect numbers? Are there infinitely many even perfect numbers? Nobody knows the answer to either of these questions. However, as we will see, we can at least tell what an even perfect number looks like: we'll prove this later, but the answer is hinted at in exercises 0.4 and 0.5.

Here's another example of a problem that is also easy to state but currently unsolved. Start with a positive integer n (your choice!); now define a new positive integer, which we will call n', as follows: if n is even, $n'=n/2$; if n is odd, $n'=3n+1$. Now, having defined n', use this same recipe to define $(n')'$, and so on. We obtain a sequence of positive integers in this manner; for example, if our initial choice was $n=7$, we obtain the sequence 7, 22, 11, 34, 17, 52, 26, 13, 40, 20, 10, 5, 16, 8, 4, 2, 1. Note that once we arrive at 1, we are essentially done, since from this point on the sequence just loops around: 1, 4, 2, 1. Let's illustrate this with another value of n, say $n=15$. We get 15, 46, 23, 70, 35, 106, 53, 160, 80, 40, 20, 10, 5, 16, 8, 4, 2, 1. So we have again arrived at 1. This suggests the question: do we *always* arrive at 1, no matter what initial choice of n was made? The assertion that we do is called the *Collatz Conjecture*; it was proposed by Lothar Collatz in 1937, but to this day nobody has either proved it or found a counterexample, although, using computers, this conjecture has been verified for literally trillions of integers. Quite recently, in September 2019, Terrence Tao announced a major breakthrough in this problem: although not completely proving the truthfulness of the conjecture, Tao did prove that for "almost all" starting values, the conjecture is at least "almost true", where "almost all" and "almost true" have rather technical mathematical meanings.

Another interesting thing happened in September 2019. For some time, mathematicians have been interested in knowing which integers can be expressed as the sum of three cubes (positive, negative or zero). Since the 1950s, computers have been used to help determine whether integers can or cannot be so expressed. The numbers 33 and 42 proved especially recalcitrant, but in September 2019 an expression of each of these numbers as a sum of three cubes was, using computers, found. The three integers whose cubes sum to 33 each have 16 digits; the three whose cubes sum to 42 have 17 digits each.

These examples illustrate that computers can play a big role in number theory problems. One way to discover patterns (that can then be proved theoretically) is to use computers to explore lots of special cases. Computers can also be used to find counterexamples to conjectures as well.

Another source of fascination over the years has been *prime numbers*. These are integers, greater than 1, that are divisible only by themselves and 1. So, for example, the first few primes are 2, 3, 5, 7 and 11. A number of questions (some easy, some hard) can be asked about these numbers and their distribution among all the positive integers. One obvious one is: is there a largest prime number? Or, putting it another way: are there only finitely many primes? The answer to this question has been known for thousands of years; it is not too difficult to prove (and we will soon see several different proofs) that there are, in fact, infinitely many primes. Quite a lot of computing time has been spent trying to discover very large primes; the largest known one, as of this writing, was discovered in December 2018 and has almost 25 million digits.

Not all questions about prime numbers are this easy to answer; there are some that are so difficult that, even after hundreds of years, their answers are not yet known. For example, note that 3 and 5, and 5 and 7, are both primes; so are 11 and 13. These are known as *twin primes* because they are two consecutive odd numbers. The *twin prime conjecture* asserts that there are infinitely many pairs of twin primes, but nobody has proved or disproved it.

In mathematics, if a problem proves very difficult, it is fairly common to look at a related but simpler problem. Since the twin prime conjecture is hard, let us consider the following weaker conjecture: that there is *some* positive integer k with the property that there are infinitely many primes differing by at most k. (This is called the *bounded gap* problem; if we could establish that $k=2$, we would have the twin prime conjecture.) For more than 100 years, this, too, was an unsolved problem, but in 2003, a mathematician named Yitang Zhang proved that such a positive integer k existed.

A number of other questions (easy to state but hard to answer) about prime numbers have been posed over the years. The famous *Goldbach conjecture* (first posed in 1742), for example, asserts that every positive even integer greater than 2 can be written as the sum of two (not necessarily distinct) primes. So, for example, $4=2+2$, $6=3+3$, $8=3+5$, $10=3+7$ (or $5+5$), $12=5+7$, etc. This conjecture has been tested and verified for literally billions of positive even integers, but nobody has yet proved that it *always* holds. What *has* been proved, however, is that every positive even integer greater than 2 can be written as the sum of a prime and another positive integer that is the product of at most two primes.

Goldbach's conjecture, if true, has an immediate consequence: every positive odd integer greater than 5 can be written as the sum of three (not necessarily distinct) primes. This latter statement, though clearly implied by Goldbach's conjecture (why?), does not itself imply it; it is therefore a weaker statement of the conjecture. Even this weaker statement, however, remained unproved for centuries, but in 2013 a proof of it was announced by Harald Helfgott.

Of course, not all problems in number theory are as easy to understand as the ones discussed above. Some are very technical—indeed, too technical to conveniently state here. An example is the famous *Riemann Hypothesis*, which is quite likely the most famous currently unsolved problem in mathematics, and one for which a prize of 1 million dollars for its solution has been offered by the Clay Mathematics Institute. (There are seven such problems, called the Millennium Prize Problems; they were proposed in 2000, and, since then, only one of them has been solved.) There are several equivalent ways to state the Riemann Hypothesis, which involves, at a very deep level, the distribution of prime numbers among all the integers.

These examples of questions in number theory don't even begin to scratch the surface of the kind of problems that arise in the subject. Numerous other examples of number theoretic questions will be discussed in the rest of the book.

One final comment: Historically, number theory was considered to be a completely "pure" subject in mathematics, without "real life" applications. The number theorist G.H. Hardy, for example, in his famous *A Mathematician's Apology*, wrote: "No one has yet discovered any warlike purpose to be served by the theory of numbers or relativity, and it seems very unlikely that anyone will do so for many years." Hardy was wrong: number theory is now recognized as having many practical applications, to warfare and other more peaceful concerns, most notably to the subject of cryptography. We will address some of these cryptographic applications later in the book.

0.1 Exercises

0.1. Find several prime numbers that can be written in the form n^2+1, for some positive integer n. The question of whether there are infinitely many such primes is also unsolved to this day. The question of whether there are infinitely many primes of the form n^3+1 is, however, easily resolved. Explain. What about primes of the form n^2-1? Explain again.

0.2. Write down, in a column, the first 25 odd prime numbers. For each one, check to see whether it can be written as the sum of two perfect squares (e.g., 3 cannot be, but 5 (= 4+1) can be). Based on the information that you gathered, formulate a conjecture as to precisely which odd primes can be so written. Do not attempt to prove this conjecture, however.

0.3. Refer back to the list of primes that you created in question 0.2 above. Check to see which primes can be written as the sum of four (or fewer) perfect squares. Again, formulate a conjecture but do not attempt to prove it.

0.4. What is the smallest perfect number that is greater than 6? (No technology allowed!)

0.5. Consider the number 6, and also the number you found in the previous exercise. Show that both numbers are of the form $2^{n-1}(2^n-1)$, where n is an integer greater than 1 and 2^n-1 is a prime. Use this observation to find yet another even perfect number. (Primes of the form 2^n-1 are called *Mersenne primes*; they will come up again in the text.)

0.6. Why does the statement "every positive odd integer greater than 5 can be written as the sum of three (not necessarily distinct) primes" follow immediately from Goldbach's conjecture?

0.7. In view of Goldbach's conjecture, one might be tempted to ask whether any integer can be written as the sum of two *composite* integers. (A composite integer is an integer, greater than 1, that is not prime.) Show that the number 11 cannot be so written. Then show that any integer that is greater than 11 *can* be so written. In doing this problem, you may assume, without proof, the basic fact that the difference of two even, or two odd, integers is even.

1

Divisibility

1.1 The Principles of Well-Ordering and Mathematical Induction

The word "number" may, depending on context, mean different things to different people. There are lots of different kinds of numbers in mathematics—real numbers, rational numbers, complex numbers, algebraic numbers and even more esoteric things like quaternions, octonions, p-adic numbers or surreal numbers. For our purposes, however, the word "number" shall, except for the last chapter in this book, refer to an element of the set $\mathbb{Z}$ of integers; i.e., the "whole numbers": ....–2, –2, 0, 1, 2, ...The reader has presumably been dealing wiwth these numbers since grade school, but probably not in any kind of theoretical sense.

We will not attempt any kind of formal definition of an integer but will instead rely on the reader's experience with them. In particular, we assume that the reader knows what an integer is; knows that there are operations of addition, subtraction and multiplication defined on the set of integers; and knows that these operations satisfy the usual rules of arithmetic: addition and multiplication, for example, satisfy the associative and commutative laws, as well as the distributive laws. We will also assume that the reader is familiar with the notion of positive and negative integers, and the basic facts concerning them (e.g., that the sum and product of two positive integers is positive).

However, it should be noted that in mathematics, if you are going to prove something about an object, you need to know precisely what that object is. Therefore, a precise approach to the study of the integers would involve writing down some axioms for these numbers and deducing things as a consequence of these axioms. To give an idea of how this is done, we have specified a set of axioms in Appendix B and shown how certain basic properties of the integers follow from them. Another Appendix (A) also provides a quick "primer" on the nature of proof and the basic principles of logic that are used in mathematics all the time.

There is one property of the integers that the reader may not have seen before, so let us single it out now. The set $\mathbb{Z}$ of integers has no smallest number; we can keep going backward in the set forever. But the set of positive integers has the smallest number 1, and therefore, our intuition tells us that in any nonempty set of positive integers we cannot regress backward infinitely

DOI: 10.1201/9781003318712-2

far. The Well-Ordering Principle, stated below (and also taken as an axiom in Appendix B), makes this intuition precise.

Well-Ordering Principle: Any nonempty set S of positive integers contains a smallest element, i.e., an element x with the property that $x \leq y$ for all $y \in$ S.

We immediately point out a trivial restatement of this principle: any nonempty set S of *nonnegative* integers has a smallest element. This is because if 0 is an element of S, then it is clearly the smallest element of it; if 0 is not an element of S, then S consists of positive integers and the Well-Ordering Principle applies.

We motivated the Well-Ordering Principle by noting that there is a smallest positive integer, namely 1. This is certainly something that most readers of this book will be happy to take on faith as a "given", based on their years of prior acquaintance with the set of integers. Yet, it is not something that was assumed as an axiom and, it turns out, can be proved easily as a consequence of the Well-Ordering Principle. Because of the simplicity of the proof, and because it illustrates how to use the Well-Ordering Principle, we take the time to prove it precisely rather than just slip it under the rug.

Theorem 1.1.1

There is no positive integer that is less than 1.

Proof. Suppose to the contrary that a positive integer less than 1 existed. Then the set S of all positive integers less than 1 is nonempty, and hence, by the Well-Ordering Principle, has a smallest element; call it x. Multiply the inequality $0 < x < 1$ by x; since we are multiplying by a positive integer, the inequality is preserved and we get $0 < x^2 < x < 1$. It follows from this that x^2 is a positive integer that is less than 1 but also less than x, which contradicts our definition of x.

This theorem finds immediate application in the next result, which states precisely, and proves, the Principle of Mathematical Induction. The reader may have already encountered this idea previously; it is a standard proof method. Because we deduce this as a consequence of the Well-Ordering Principle, we label it as a theorem.

Theorem 1.1.2

(Principle of Mathematical Induction) Suppose that

- S is a subset of the set of positive integers,
- $1 \in$ S, and
- $n+1 \in$ S whenever $n \in$ S.

Then S consists of all positive integers.

Proof. Assume, hoping for a contradiction, that there is a positive integer that is *not* in S. Then, by the Well-Ordering Principle (applied to the nonempty set of all such integers) there must be a *smallest* positive integer not in S; call it k. Note that $k \neq 1$ (because $1 \in S$), so $k-1$ is a positive integer (note that we are using Theorem 1.1.1 here!) and because it is smaller than k, must be in S. However, by assumption, since $k-1 \in S$, it must be the case that $k = (k-1)+1 \in S$, a contradiction. This contradiction yields the desired result.

The Principle of Mathematical Induction is typically used as a proof tool. If asked to prove that a certain statement is true for all positive integers n, one first proves it is true for $n=1$ and then, assuming it is true for n, proves it true for $n+1$. In the language of the preceding theorem, we let S be the set of all positive integers n for which the result is true; the proofs just discussed then show that S is the set of all positive integers, and we are done.

We illustrate this method with a simple example: we will prove that the sum of the first n positive integers is equal to $n(n+1)/2$. For $n=1$, this is obvious because $(1 \times 2)/2 = 1$. So, now assume the result is true for n, and let us examine the sum of the first $n+1$ positive integers; we want to prove this is equal to $(n+1)(n+2)/2$. Well,

$$1+2+....+n+(n+1) =$$

$$(1+2+...+n)+(n+1) =$$

$$n(n+1)/2+(n+1) =$$

$$n(n+1)/2+2(n+1)/2 =$$

$$(n(n+1) +2\ (n+1))\ /2 =$$

$(n+1)(n+2)/2$, completing the proof.

The Principle of Mathematical Induction has alternative forms. One is the Strong Induction Principle, which we state below. The proof is similar to that of Theorem 1.1.2 and is therefore omitted.

Theorem 1.1.3 (Strong Induction Principle)

Suppose that

- S is a subset of the set of positive integers,
- $1 \in S$, and
- $n+1 \in S$ whenever $1, 2, ..., n \in S$, for any positive integer n.

Then S consists of all positive integers.

To use the Strong Induction Principle in a proof, we therefore first prove that 1 is in S, and then we assume, for an arbitrary positive integer n, that S contains all the positive integers from 1 to n, and we use that to prove that $n+1$ $\in$ S. As an example of a result that can be easily proved using Strong Induction but not so easily proved using "regular" induction, there is the theorem that any integer greater than 1 is either a prime number or a product of prime integers. This is part of the Fundamental Theorem of Arithmetic, which is proved in Section 1.5 of this chapter, after prime integers have been defined. The proof we give actually uses Well Ordering, but Strong Induction can also be used in the proof.

Another variant on the Principle of Mathematical Induction starts the induction with a positive integer k rather than 1. In other words, instead of assuming $1 \in S$, we assume $k \in S$, and that $n+1 \in S$ whenever $n \in S$. This allows the conclusion that S contains all integers that are greater than or equal to k. We leave to the reader a precise formulation and proof of this result.

From this point on, we will freely use all the basic facts about addition, multiplication and ordering on the set of integers that the reader has been using for years. We will not attempt to prove, for example, that if r is any real number then there is an integer n that is greater than r, and also an integer m that is less than r.

Exercises

1.1   Prove by induction that the sum of the first n positive odd integers is equal to n^2.

1.2   Prove by induction that $2^n > n$ for every positive integer n.

1.3   Prove by induction that a set with n elements has 2^n subsets.

1.4   If n is a nonnegative integer, define the nth *Fermat number* F_n to be $2^{2^n} + 1$. Use mathematical induction to prove that, for every n, $F_0 \ldots F_n + 2 = F_{n+1}$. (We will see Fermat numbers again. They pop up in unexpected places in mathematics, including the geometric question of when a regular polygon with n sides can be constructed with compass and straightedge alone.)

1.5   Find the error in this fallacious "proof" that all billiard balls have the same color: "We will prove that, for any positive integer n, in any set of n billiard balls, they all have the same color. This is obvious if $n=1$. Now assume the result is true for n and consider a set of $n+1$ billiard balls; let us denote them by $B_1, \ldots, B_{n+1}$. By our inductive assumption, all the balls in the set of n balls $\{B_1, \ldots, B_n\}$ have the same color; without loss of generality, let us say the color is black. Now consider the set $\{B_2, \ldots, B_{n+1}\}$. This is also a set of n balls and so they must have the same color as well. But this color must be black because B_2 is in both sets. So all the billiard balls $B_1, \ldots, B_{n+1}$ have the same color, and we are done."

1.6 In this book, we assumed the Well-Ordering Principle as an axiom and used it to prove the principle of mathematical induction as a theorem. Prove that we could have done things in reverse and deduced the principle of well-ordering as a theorem if we assumed the principle of mathematical induction as a theorem. (We summarize this state of affairs by saying that these two principles are *equivalent*.)

1.7 Let S be a set of integers, every one of which is greater than or equal to a fixed integer k. Prove that S contains a smallest element. (We do not, of course, assume that k is an element of S.)

1.8 Let k be an integer. Prove that there is no integer n such that $k<n<k+1$.

1.2 Basic Properties of Divisibility

If m and n are integers, we say that m divides n (denoted $m \mid n$) if there is an integer k such that $n=km$. Other ways to say this are "n is divisible by m" or "n is a multiple of m". Intuitively, this means that "m goes evenly into n". An integer that is divisible by 2 is called *even*; an integer that is not even is called *odd*. The *parity* of an integer refers to its "evenness" or "oddness".

The following theorem collects some of the very basic properties of the divisibility relation. The proofs of these properties are quite simple and provide good practice in writing straightforward proofs; for this reason, the proofs are (with the exception of part (f)) given in Appendix A, on proof-writing. We will use these basic results constantly in the rest of the book, often without explicit mention.

Theorem 1.2.1

If m, n and r are integers, then the following are true:

(a) $n \mid n$ (every integer divides itself)
(b) $1 \mid m$ (1 divides every integer)
(c) if $n \mid m$ and $m \mid r$ then $n \mid r$ (transitivity of divisibility)
(d) if $n \mid m$ and $n \mid r$ then $n \mid m+r$ and $n \mid m-r$ (addition and subtraction)
(e) if $n \mid 1$ then $n=\pm 1$ (divisors of 1)
(f) if $n \mid m$ and $m \mid n$ then $n=\pm m$ (anti-symmetry)

Another way of saying that $m \mid n$ is to say that n leaves a remainder of 0 when divided by m. Although we can't expect this to happen all the time, it is true (as you no doubt know from grade school) that we can divide an integer a by a positive integer b to obtain a quotient q and remainder r, with r being a non-negative integer less than b. The precise statement of this result is called the Division Algorithm (a slight misnomer, since it is not really an "algorithm" in the usual sense). Although we stated earlier that we will simply assume as known all the basic facts about integer addition and multiplication, this fact involves division and, we think, is best proved, particularly because the proof provides a nice use of the Well-Ordering Principle.

Theorem 1.2.2

(Division Algorithm) If a and b are integers, with b positive, then there exist unique integers q and r such that $a=bq+r$ and $0 \le r < b$.

Proof. We first prove that such q and r exist and will then prove that they are unique. First note that if $b \mid a$, then existence is obvious (with $r=0$). So we assume without loss of generality (to prove existence) that b does not divide a. Let S be the set of all positive integers of the form $a-bx$, as x ranges over the integers. It is obvious that S is nonempty; any x that is less than $\frac{a}{b}$ will give an element of S. So, by Well-Ordering, S contains a smallest element, which we will call r. By definition of r, it can be written as $a-bq$ for some integer q. So we know $a=bq+r$; it therefore suffices to prove $0 \le r < b$, and since we know r is positive, we need only show that $r < b$. We can't have $r=b$ because then b would divide a, contrary to our assumption. (Why?) If we had $r > b$, then $a-b(q+1)=r-b$ would be positive, less than r, and in the set S, a contradiction. So in fact $r < b$, and we have shown the existence of a quotient and remainder satisfying the conditions of the theorem.

We next prove uniqueness. Specifically, we prove that if $a=bq_1+r_1=bq_2+r_2$, with both r_1 and r_2 nonnegative and less than b, then $r_1=r_2$ and $q_1=q_2$. It suffices to prove that $r_1=r_2$, as simple algebra then would establish $q_1=q_2$. We use a proof by contradiction, and assume instead, without loss of generality, that $r_1 > r_2$. We then have, by some more simple algebra, $r_1 - r_2=b(q_1 - q_2)$. Since b and $r_1 - r_2$ are both positive, $q_1 - q_2$ must be positive as well, which would then imply that the right-hand side of this equation is at least equal to b. But this is a contradiction, because $r_1 - r_2 \le r_1 < b$. This concludes the proof.

The division algorithm is a powerful tool in number theory, and we will begin putting it to use in the very next section, where we discuss the greatest common divisor of two integers. For the moment, though, we record an easy but important consequence of it. If $b=2$, then the division algorithm tells us that any integer a can be written as either $2q$ or $2q+1$, but not as both. Integers of the first type are (as we mentioned above) called *even*, and integers of the second type are called *odd*. The uniqueness of the quotient and remainder in the Division Algorithm tells us that no integer can be both even and odd. This surely does not come as a surprise to you, but it's nice to know it "officially". It is easy to see that the product of two odd integers is odd (see the exercises). In particular, if n is an integer and n^2 is even, then n must be even. Just these basic facts allow us to prove a result of immense historical significance, namely that $\sqrt{2}$ is irrational. (Recall that a rational number is a fraction; more precisely, it is a number of the form $\frac{a}{b}$, where a and b are integers.)

Our proof of the irrationality of $\sqrt{2}$ gives a preview of the concept of greatest common divisor, discussed in the next section.

Theorem 1.2.3

$\sqrt{2}$ is irrational.

Proof. Suppose, hoping for a contradiction, that we could write $\sqrt{2} = \frac{a}{b}$, where a and b are positive integers. We can and do assume this fraction is in "lowest terms", meaning that a and b have no positive divisors in common except 1. (We can always divide both a and b by any common divisors greater than 1 without changing the value of the fraction.) Squaring both sides and clearing denominators gives $2b^2 = a^2$. This equation tells us that a^2 is even, and hence, by the observation above, this means that a is even, say $a = 2k$. Squaring and substituting gives $2b^2 = 4k^2$ or $b^2 = 2k^2$, which implies that b^2, and hence b, is even. But if a and b are both even, then the fraction $\frac{a}{b}$ is not in lowest terms, a contradiction.

The historical importance of this result can be traced back to at least as far as the ancient Greeks, who believed that any two lengths were commensurable, i.e., both a multiple of some other length. If that were true, though, then it would be true for both the side of a unit square and its diagonal (which by the Pythagorean theorem has length $\sqrt{2}$) and that would mean that $\sqrt{2}$ is rational. The realization that the side and diagonal of a square are not commensurable had important historical consequences and led Greek geometers to separate the concepts of number and segment and to develop an intricate "theory of proportions" to deal with these issues.

The Division Algorithm also underlies the idea of writing integers to "different bases". This is discussed in Section 1.6 at the end of this chapter, but this section can be read now if desired.

Exercises

1.9 Prove or disprove: if a, b, c are integers and $a \mid bc$, then $a \mid b$ or $a \mid c$.

1.10 Prove or disprove: if a, b, c are integers and $a \mid b + c$, then $a \mid b$ or $a \mid c$.

1.11 Prove part (f) of Theorem 1.2.1.

1.12 Prove that the product of two odd integers is odd.

1.13 Prove that the sum of two odd integers is even.

1.14 Prove that the product of any three consecutive integers is divisible by 3. Generalize.

1.3 The Greatest Common Divisor

Suppose that a and b are integers, not both 0. Then there is at least one positive integer, namely 1, that divides both a and b—i.e., is a common divisor of these two integers. On the other hand, since a nonzero integer has only a finite number of divisors, there must be a finite number of common divisors of a and b, and, hence, a largest one. This greatest common divisor (gcd) turns out to be very useful, and in this section of the text, we will explore it in more depth. But first, we will give a somewhat different (but equivalent) definition of the greatest common divisor. We could simply define it to be, literally, the largest of all common divisors; an advantage to doing it this way would be that the existence of the greatest common divisor would then obvious. But a disadvantage is that we would have to prove certain properties about the gcd. We will, therefore, give a different definition, one which incorporates these properties in the definition itself. We will then have to *prove* the existence of the gcd, but, having done so, we will not only know that the gcd exists, but that it satisfies certain properties.

<u>Definition 1.3.1</u> (Greatest Common Divisor)

If a and b are integers, not both 0, then the greatest common divisor of a and b, denoted $\gcd(a, b)$, is the positive integer d with the following properties:

   (i) d divides both a and b
   (ii) if k is any other integer that divides a and b, then k divides d

So, the gcd of a and b is not only *greater* than or equal to any other common divisor of these integers, it is a *multiple* of that common divisor. For example, if $a=8$ and $b=20$, then the only positive common divisors of a and b are 1, 2 and 4; thus, $\gcd(8, 20)=4$, which is a multiple of the only other common divisors, 1 and 2.

As noted above, it is not immediately obvious that $\gcd(a, b)$ always exists. What is clear (see the exercises) is that if the gcd of a and b exists, it is unique. (This is why we refer in the definition to "*the* positive integer d".) We will prove the existence—and, simultaneously, a useful property about the gcd. The proof that we will present uses a concept called *ideals*. This may seem like the long way around the barn, but ideals are useful in more general algebraic settings (see, for example, Chapter 6 for a hint of this), and so it's not a bad idea to first see them defined for the integers.

Definition 1.3.2 (Ideals)

A subset I of $\mathbb{Z}$ is called an *ideal* if

(i)   0 is an element of I
(ii) if a and b are elements of I, so is $a+b$ (closure under addition)
(iii) if a is an element of I and b is any integer whatsoever, ab is an element of I ("super-closure" under multiplication)

Note the difference between conditions (ii) and (iii): condition (ii) is ordinary closure under addition (that if two integers are both in I, then so is their sum), but condition (iii) is a stronger condition (hence the ad-hoc term "super-closure", which is not standard terminology): not only is I closed under multiplication in the ordinary sense, but I contains any product where *even one* of the terms in the product is in I.

Examples of ideals are easy to give: the two "trivial" ones are {0} and $\mathbb{Z}$ itself. For a less trivial example, let k be any integer and denote by *<k>* the set of all multiples km of k. It is easy to verify that this is an ideal, called the *principal ideal generated by k*. Observe that this example includes the two previous ones as special cases, since clearly $\{0\} = <0>$ and $\mathbb{Z} = <1>$.

As it happens, the principal ideals in $\mathbb{Z}$ are *all* the ideals of $\mathbb{Z}$.

Theorem 1.3.3

If I is any ideal of $\mathbb{Z}$, then $I = <b>$ for some integer b.

Proof. If $I = \{0\}$ then, as just observed, $I = <0>$, and we are done. So suppose I contains at least one nonzero integer k. Then I also contains $-k\ (=(-1)k)$ by property (iii). At least one of k or $-k$ is strictly positive, so I contains a positive integer, and hence, by the Well-Ordering Principle, I contains a smallest positive integer, say b. We claim that $I = <b>$. Clearly, by "super-closure" of multiplication, $<b> \subseteq I$. For the reverse inclusion, let a be an arbitrary element of I. By the Division Algorithm, we may write $a = bq + r$, where $0 \leq r < b$. Because both a and b are in I and I is an ideal, it follows that $r = a - bq$ is also in I. If $r \neq 0$, then this would give us a positive element of I that is strictly smaller than b, a contradiction. Thus $r = 0$, and so $a = bq \in <b>$ finishing the proof.

This result has significance in abstract algebra: it says, intuitively, that an algebraic system in which we have an analog of the Division Algorithm is one in which every ideal is principal. In more advanced courses, we would phrase this as "every Euclidean domain is a Principal Ideal Domain".

We now use this result to prove the existence of a gcd of two integers and also to prove, at the same time, an additional fact about that gcd. We say that an integer k is a *linear combination* of a and b if we can write $k = ax + by$ for some integers x and y. Using this terminology, we now prove:

Theorem 1.3.4

If a *and* b are integers, not both zero, then the gcd of a *and* b exists and is a linear combination of a and b.

Proof. Let $I = \{ax + by : x, y \in \mathbb{Z}\}$ be the set of all linear combinations of a and b. Note that both $a = a1 + b0$ and $b = a0 + b1$ are in I, and at least one of these integers is nonzero. Since it is also easy to see (check this!) that I is an ideal, I is principal by the previous theorem and therefore consists precisely of the multiples of some integer d. It follows from our previous remarks that we can assume that $d > 0$ (why?). We will prove that d is the gcd of a and b. Since it is obvious that d is also a linear combination of a and b by the way it is defined, this will complete the proof.

To show that the positive integer d is the gcd of a and b, we first observe that d divides both a and b. This follows from the observation, made in the previous paragraph, that I contains both a and b, and every element of I is a multiple of d by the way it is defined.

Finally, suppose k also divides both a and b. Then it is clear that k also divides any linear combination of a and b. But one such linear combination is d itself. So $k \mid d$, and this completes the proof.

Although this theorem guarantees the existence of the greatest common divisor, the proof given above does not provide a method for actually finding it in a particular case. In the next section, however, we will discuss a useful algorithm for computing the gcd of two integers and expressing it as a linear combination of these integers. For the moment, we content ourselves with a simple example: suppose we wish to find the gcd of 114 and 102. We can begin by listing the divisors of 102: 1, 2, 3, 6, 17, 51 and 102. Of these integers, we see that 1, 2, 3 and 6 also divide 114. So 6 is the gcd, and, consistent with the theorem, all the other common divisors (1, 2 and 3) divide it.

Two integers whose greatest common divisor is equal to 1 are called *relatively prime*. If integers a and b are relatively prime, then the previous result establishes that 1 can be written as a linear combination of a and b. It is quite easy to see that the converse of this result is also true.

Theorem 1.3.5

Two integers a and b, not both 0, are relatively prime if and only if the equation $ax+by=1$ has a solution in integers x and y.

Proof. As noted, if gcd $(a, b)=1$, then Theorem 1.3.4 guarantees the existence of integers x and y satisfying $ax+by=1$. For the converse, suppose this equation has a solution in integers x and y. Let d denote the gcd of a and b. Then by our basic list of properties of divisibility, it is clear that d divides $ax+by$. But this means that d divides 1, which (since d is positive) implies that $d=1$. Hence, a and b are relatively prime.

Students frequently misinterpret the previous theorem and misread it as saying that "a and b have gcd n if and only if the equation $ax+by=n$ has a solution in integers x and y". This is false: the equation $2x+3y=2$ has the obvious solution $x=1$, $y=0$ but the gcd of 3 and 2 is not 2; it is 1. The true state of affairs is given by the following theorem, the proof of which is similar to that of Theorem 1.3.5, and which includes Theorem 1.3.5 as a special case (because the only positive integer that divides 1 is 1).

Theorem 1.3.6

For two integers a and b, not both 0, the equation $ax+by=n$ has a solution in integers x and y if and only if the gcd of a and b divides n.

Proof. Exercise.

It is worth noting that the integers x and y in the previous theorem are not unique. Indeed, suppose that the equation $ax+by=n$ has a solution (X, Y). Let d denote the greatest common divisor of a and b, and consider the integers $x'=X+(b/d)t$ and $y'=Y-(a/d)t$, where t is any integer whatsoever. It is easy to show by direct calculation that x' and y' are also solutions to the equation $ax+by=n$. We will finish this section by proving that these are *all* the solutions—i.e., that any solution is of this form, for some integer t. We first need some preliminary results that are important in their own right.

Theorem 1.3.7

If a, b and c are integers, $a \mid bc$, and a and b are relatively prime, then $a \mid c$.

Proof. Since a and b are relatively prime, we can express 1 as a linear combination of them: $1=ax+by$. Multiplying both sides of this equation by c gives $c=acx+bcy$. Now notice that because $a \mid bc$, a divides the second summand on the right-hand side of this equation; it also obviously divides the first summand, and hence it divides their sum, which is c.

Theorem 1.3.8

If a, b and c are integers, $a \mid c$, $b \mid c$ and a and b are relatively prime, then $ab \mid c$.

Proof. We know that $c=ax$ and $c=by$ for some integers x and y. So $ax=by$, and hence a divides by. But by the previous theorem, this means a divides y. Thus, we can write $y=az$ for some integer z. Thus $c=by=abz$, which makes it obvious that $ab \mid c$.

An induction argument applied to this theorem gives this extension:

Theorem 1.3.9

If $m_1, \ldots, m_n$ are integers, any two of which are relatively prime, and each m_i divides an integer c, then the product $m_1 \ldots m_n$ also divides c.

Proof. Induction on n. We defer the proof, however, to Section 1.5, after we have developed some more machinery. (The reader might wish to try proving this now; what problem arises?)

We can now, as promised, finish up the discussion of roots of the equation $ax+by=n$.

Theorem 1.3.10

Consider the equation $ax+by=n$ where a and b are integers, not both 0. Suppose that (X, Y) is a solution to this equation. Let d denote the greatest common divisor of a and b. If (x', y') is any solution to this equation, then, for some integer t, $x'=X+\dfrac{b}{d} t$ and $y'=Y-\dfrac{a}{d} t$.

Proof. Assume a is nonzero. (Either a or b is, so we can assume without loss of generality that a is.) We know that $ax'+by'=n=aX+bY$. Simple algebra then yields $\dfrac{a}{d} (x' - X)= \dfrac{b}{d} (Y-y')$, from which it follows that $\dfrac{a}{d}$ divides $\dfrac{b}{d} (Y-y')$. By Theorem 1.3.7 and Exercise 1.19, it follows that $\dfrac{a}{d}$ divides $(Y-y')$. Hence, for some t, $(Y-y')= \dfrac{a}{d} t$. Solving for y 'gives $y'=Y- \dfrac{a}{d} t$. Substituting this in $\dfrac{a}{d} (x' - X)= \dfrac{b}{d} (Y-y')$ gives the desired formula for x'.

We illustrate this theorem with a simple example. Suppose $a=5$ and $b=7$. Then of course $d=1$, and simple arithmetic tells us that the equation $5x+7y=1$ has the solution $X=-4$, $Y=3$. So, by the theorem, the general solution to $5x+7y=1$ is $(-4+7t, 3-5t)$. If, for example, we take $t=2$, this yields the solution $(10, -7)$, and mental arithmetic shows that this is, indeed, a solution.

Exercises

1.15 Give an example of a nonempty set of integers that is closed under addition but is not an ideal. Give an example of a nonempty set of integers that is "superclosed" under multiplication but is not an ideal.

1.16 Show that the intersection of two ideals in the set of integers is always an ideal, but the union need not be.

1.17 Generalizing the previous exercise, prove that, in fact, the only time that $I \cup J$ is an ideal if either I or J is a subset of the other.

1.18 Find, with proof, all ideals of the integers that contain the integer 1.

1.19 Prove that if $d = \gcd(a, b)$, then a/d and b/d are relatively prime.

1.20 If n is a nonzero integer, what is $\gcd(n, 0)$? Prove your answer.

1.21 If n is an integer, what is $\gcd(n, n+2)$? Prove your answer (which will depend on the parity of n).

1.22 If a and b are relatively prime integers, what is the gcd of $2a$ and $2b$? Prove your answer, and then state and prove a generalization of this result.

1.23 If a and b are relatively prime integers, what are the possible values of $\gcd(a+b, a-b)$? Prove your answer in detail.

1.24 Show by example that the conclusions of Theorems 1.3.7 and 1.3.8 are not true if we do not assume that a and b are relatively prime.

1.25 Prove that an integer can be expressed as the difference of two squares if and only if it is odd or divisible by 4.

1.26 If a and b are positive integers, we define the least common multiple of a and b to be the smallest positive integer m with the property that both a and b divide m. Prove that m exists, and in fact $m = ab/d$, where d denotes the gcd of a and b.

1.4 The Euclidean Algorithm

In this section, we discuss a computational method for computing the greatest common divisor d of two integers a and b; the method also allows us to find integers x and y such that $d = ax + by$. The key to this method (which was known to Euclid, more than 2000 years ago) is the following theorem.

Theorem 1.4.1

Suppose a and b are nonzero integers, and $a = bq + r$. Then gcd $(a, b) = $ gcd (b, r).

Proof. Let us denote gcd (a, b) by d. We will show that d is also the greatest common divisor of b and r by showing that it satisfies the defining properties of that greatest common divisor. We already know that d is positive, so it suffices to show that d divides both b and r, and also that d is a multiple of any other divisor of b and r. Both of these, however, follow from our basic properties of divisibility. Since d divides both a and b, it divides $a - bq = r$. So d divides b and r. Also, if k is any divisor of b and r, then k also divides $bq + r$, which is a. As a divisor of a and b, k divides d.

We will use this theorem in a second to see that the algorithm works. But first, we must describe the algorithm.

Euclidean Algorithm. Given two integers a and b with, say, $a > b$, follow these steps to compute the greatest common divisor of a and b:

- First, apply the Division Algorithm to a and b, getting quotient q_1 and remainder r_1
- Then, apply the Division Algorithm to b and r_1, getting quotient q_2 and remainder r_2
- Next, apply the Division Algorithm to r_1 and r_2, getting quotient q_3 and remainder r_3
- Repeat this process until we get a remainder of 0
- The last nonzero remainder is the greatest common divisor of a and b

In symbols, we have the following chain of equations:

$$a = bq_1 + r_1$$

$$b = r_1 q_2 + r_2$$

$$r_1 = r_2 q_3 + r_3$$
$$\vdots$$
$$r_n = r_{n+1} q_{n+2}$$

which results in the conclusion that r_{n+1} is the greatest common divisor of a and b.

Before proceeding further, we should perhaps call attention to a possible issue: our statement of the algorithm used the phrase "until we get a remainder of 0". Must we, in fact, *always* eventually get a zero remainder? The answer is yes: note that by construction of the integers r_i, we get a strictly decreasing sequence of nonnegative integers: the second equation above gives $r_2 < r_1$, the third gives $r_3 < r_2$, etc. A strictly decreasing sequence of nonnegative integers can't go on forever; it must eventually reach 0.

We will shortly prove that the Euclidean Algorithm does, in fact, produce the gcd of a and b, but before doing so, it would be helpful to work through a specific example. Suppose, for example, that we want to compute the gcd of $a=824$ and $b=260$. We perform the following calculations:

$$824=260(3)+44$$

$$260=44(5)+40$$

$$44=40(1)+4$$

$$40=4(10)$$

which produces 4 (the last nonzero remainder) as the gcd of 824 and 260.

To see that this method works in general, refer back to the general system of equations listed above, and note that, by Theorem 1.4.1, we have:

gcd $(a, b)=$gcd $(b, r_1)=$gcd $(r_1, r_2)=\ldots..$ gcd $(r_n, r_{n+1})=$gcd $(r_{n+1}, 0)=r_{n+1}$, where, for the last equality, we used the (very simple) result of Exercise 1.20.

Because 4 is the greatest common divisor of 824 and 260, we know that it can be expressed as a linear combination of these two integers: i.e., there exist integers x and y such that $4=824x+260y$. The calculations in the Euclidean Algorithm allow us, by proceeding backward, to actually find integers x and y that work. The trick is to start at the penultimate equation, solve for 4 as a linear combination of the previous remainders and keep "backward solving" until we arrive at a linear combination of the original integers. Observe:

$$4=44 - 40(1)$$

$$= 44 - (260 - 44(5))$$

$$= 44(6) - 260$$

$$= [824 - 260(3)](6) - 260$$

$$= 824(6) - 260(18) - 260$$

$$= 824(6) - 260(19).$$

Thus, we have expressed 4 as a linear combination of 824 and 260. (The prudent thing to do, of course, would be to check the right-hand side with a calculator to make sure that it gives us the value 4; it does.)

Let's try one more example. Suppose we want to find the gcd of $a=234$ and $b=63$. We get

$$234=63(3)+45$$

$$63 = 45(1) + 18$$

$$45 = 18(2) + 9$$

$$18 = 9(2)$$

so the gcd is 9. Working backward to express 9 as a linear combination of 234 and 63, we get:

$$9 = 45 - 18(2)$$

$$= 45 - (63 - 45)\,(2)$$

$$= 63(-2) + 45(3)$$

$$= 63(-2) + [234 - 63(3)](3)$$

$$= 63(-11) + 234(3),$$

which we can also check to be correct. (The right-hand side is $-693 + 702$.)

We close this section by stating, without proof, a result that might be of interest to readers who care about the computational complexity of algorithms. It is known as *Lame's theorem*.

Theorem 1.4.2

The number of steps required in the Euclidean Algorithm is less than or equal to five times the number of digits in the smaller of the two integers whose gcd is being determined.

Exercises

1.27 Use the Euclidean Algorithm to find the gcd of 1024 and 342. Then express this gcd as a linear combination of 1024 and 342.

1.28 Same directions as the previous question, but this time with the integers 490 and 102.

1.29 The famous *Fibonacci sequence* F_n is defined inductively as follows: $F_0 = 0$, $F_1 = 1$, and, for nonnegative n, $F_{n+2} = F_{n+1} + F_n$. (So, for example, the first few terms in the sequence are 0, 1, 1, 2, 3, 5, 8, 13 and so on.) Apply the Euclidean Algorithm to prove that any two consecutive terms in the Fibonacci sequence are relatively prime.

that $p_1 \mid q_1 q_2 \ldots q_n$, which by Theorem 1.4.3 (the generalized Euclid's Lemma) implies that $p_1 \mid q_i$ for some i. Relabeling if necessary, we may assume that $i = 1$. But if $p_1 \mid q_1$, then $p_1 = q_1$. Thus, we can divide both sides of the equation $p_1 p_2 \ldots p_m = q_1 q_2 \ldots q_n$ by p_1, getting the equation $p_2 \ldots p_m = q_2 \ldots q_n$.

The idea now, of course, is to keep this process up, equating (at each stage) one of the primes p_i with the (relabeled if necessary) prime q_j. If we knew that $m = n$, then this would prove the result. But in fact we do know this: if, for example, we had $m > n$, then we would eventually arrive at an equation where the left-hand side was a product of one or more p_i and the right-hand side was 1. This would imply $p_m \mid 1$, an obvious contradiction. A similar contradiction would occur if we assumed that $n > m$. Thus $m = n$, and the proof is complete.

It follows from the preceding result that any integer n greater than 1 can be expressed uniquely in the form $\prod p_i^{a_i}$ where the symbol $\prod$ denotes "product", each p_i is a prime, and each a_i is a positive integer: just collect all the primes together that appear in the unique factorization of n. However, there is currently no known computationally feasible method for *finding* this prime factorization. This is a fact that, as we will see, is at the heart of a lot of cryptography theory.

As a consequence of unique factorization, note that if $n = \prod p_i^{a_i}$ and $m \mid n$ for some nonnegative integer m, then we can write m as $\prod p_i^{b_i}$ where $0 \leq b_i \leq a_i$. This is because any prime that appears in the factorization of m must (by uniqueness of prime factorization) be one of the primes that appears in the factorization of n; if a prime appears in the factorization of n but not in m, we can think of it as appearing with exponent 0. So, in particular, if $n = m^k$ for some nonnegative integer m, then (again by uniqueness) $a_i = k b_i$. The converse is also easy to see. So, we have proved.

Theorem 1.5.5

The integer $n = \prod p_i^{a_i}$ is a kth power if and only if every exponent a_i is divisible by k. In particular, a positive integer n is a square if, and only if, in the prime factorization of n, every prime appears an even number of times.

This theorem, in turn, allows us to give an easy proof of the next theorem.

Theorem 1.5.6

If m and n are positive relatively prime integers, then mn is a kth power if and only if both m and n are.

Proof. Exercise.

As our final result about primes in this section, we use the fact that any integer greater than 1 has a prime divisor to give a famous proof of the fact that there are infinitely many primes. This proof was known to Euclid; it appears (in less modern terminology, of course) in his famous treatise *The Elements*, which dates back to roughly 300 B.C. It is a masterpiece of mathematical reasoning: short, beautifully elegant, easy to understand and insightful. Every mathematics major should see this proof before he or she graduates from college.

Theorem 1.5.7

There are infinitely many primes.

Proof. Suppose to the contrary that there are only a finite number of primes and denote them $p_1, p_2, ..., p_n$. Now let $N = p_1 p_2 ... p_{n+1}$. The number N is obviously greater than 1, and so, by the Fundamental Theorem of Arithmetic, has a prime divisor. But, by assumption, this prime divisor must be one of the primes $p_1, p_2, ..., p_n$, as these are the only primes that exist. Denote this prime divisor by p_i. Now observe that $p_i \mid N$ and $p_i \mid p_1 p_2 ... p_n$, so $p_i \mid N - p_1 p_2 ... p_n$. But this says $p_i \mid 1$, an obvious contradiction.

There are a great many other proofs of the infinitude of primes, a few of which appear in the exercises at the end of this section. It should also be noted that this result can be strengthened in several respects. For example, for readers who are familiar with infinite series, we have the following result (which we state without proof) that clearly implies that the number of primes cannot be finite.

Theorem 1.5.8

The series $\sum 1/p$ of reciprocals of the prime integers diverges.

We close this section by stating two other results, the proofs of which are beyond the scope of this text. We do ask the reader, however, to prove a few special cases of the first theorem as exercises.

Theorem 1.5.9

(Dirichlet's theorem on primes in arithmetic progression) If a and b are two relatively prime positive integers, then there are infinitely many primes of the form $an+b$.

Our second result has the misleading name Bertrand's Postulate; however, it is not a postulate (or axiom) but actually a theorem, the proof of which is likewise beyond the scope of this text. The reason for the name is historical: this result was first conjectured by Bertrand in 1845 and proved by Chebyshev in 1850.

Theorem 1.5.10

(Bertrand's Postulate) If $n > 1$ is any positive integer, then there is a prime p satisfying $n < p < 2n$.

Exercises

1.30 If n is a positive integer, define n *factorial* (written $n!$) to be the product of all positive integers from 1 to n. Thus, for example, $3! = 6$ and $4! = 24$. Prove, by considering $n! + 1$, that if n is any positive integer, there is a prime greater than n. (This, of course, gives another proof of the infinitude of primes.)

1.31 If p and q are distinct odd primes, find, with proof, the gcd of $p + q$ and $p - q$.

1.32 Prove that there are infinitely many primes of the form $4n - 1$.

1.33 Verify Bertrand's Postulate for every integer n between 2 and 20, inclusive.

1.34 Prove Theorem 1.5.3.

1.35 Prove the existence part of Theorem 1.5.4 by Strong Induction.

1.36 If p and q are distinct primes, find, with proof, the gcd of pq and $p + q$.

1.37 Prove Theorem 1.5.6.

1.38 Find all primes p with the property that p, $p + 2$ and $p + 4$ are all primes. (Hint: use the Division Algorithm, and see what the possible remainders are when p is divided by 3.) So, while the study of twin primes may be difficult (see Chapter 0), the number of prime *triples* is thoroughly understood.

1.39 This problem describes a method for finding all primes less than or equal to some positive integer n. It is called the *Sieve of Erathosthenes*. The problem is to prove this method works. Here is the method: List all the integers from 2 to n, inclusive. Then, start with the first integer in the list, 2, and cross out all integers in the list that are divisible by 2, other than 2 itself. Now start with the first integer remaining, 3, and cross out all integers in the list divisible by 3 except 3 itself. Keep doing this until the first surviving entry in the list is greater than or equal to $\sqrt{n}$. All the remaining entries in the list are prime, and these are all the primes that are less than or equal to n.

1.40 Apply the Sieve of Erathosthenes to find all prime integers that are less than or equal to 50.

1.41 If you haven't already done so, please read the discussion of Goldbach's Conjecture in Chapter 0. Then prove this related-sounding (but *vastly* easier!) result: there are infinitely many positive odd integers that cannot be written as the sum of two primes.

1.42 Suppose that p is a prime, and that $p \mid a^k$, where a and k are positive integers. Prove that $p^k \mid a^k$.

1.43 Suppose that m and n are two positive integers, whose prime factorizations are $m = \prod p_i^{a_i}$ and $n = \prod p_i^{b_i}$. (We can assume the primes are the same in both factorizations because we allow the exponents to be 0.) Explain how to find the greatest common divisor in terms of these prime factorizations. Why does this not give us a computationally efficient method for finding the gcd of two integers?

1.6 Numbers to Different Bases

Our system of writing multi-digit nonnegative numbers is a *decimal* system, based on powers of 10: the number 472, for example, is a shorthand way of writing $4 \cdot 10^2 + 7 \cdot 10^1 + 2 \cdot 10^0$. Other than the fact that most people have ten fingers, however, there is nothing magical about the number 10, and we can use any integer b greater than 1 as a "base". The precise statement of the theorem we have in mind is:

Theorem 1.6.1

Let $b > 1$ be any integer. Then any nonnegative integer m can be written uniquely as a sum $a_n b^n + \ldots + a_1 b + a_0$, where each a_i satisfies $0 \le a_i < b$.

We won't prove this result because we won't really use it in the remainder of this book, and the proof is rather dry and technical. However, it is worth noting that the proof really amounts to repeated use of the Division Algorithm, so we will give a few examples illustrating this fact.

Before giving these examples, though, a few brief remarks are in order. First, the case $b = 10$ just gives the ordinary decimal expansion of a number, where each a_i can be an integer from 0 to 9. Also, when the number m is written in the form specified above, it is traditional to write $m = (a_n a_{n-1} \ldots a_0)_b$, and we say that m has been written in the base b. Finally, when $b = 2$, each a_i is either 0 or 1, so when we write m out in base 2 we just get a sum of distinct powers of 2. In other words, every coefficient in the base 2 expansion of an integer is 0 or 1. Computer science majors will no doubt recognize the significance of writing any nonnegative integer as a string of 0s and 1s.

We now illustrate how to write a nonnegative integer in base b and also illustrate how the Division Algorithm is related to this idea. Suppose, for example, that we want to write the number 1000 in base 2. We first determine the largest power of 2 that is less than or equal to 1000; in this case, that is clearly $512=2^9$. So we divide 1000 by 512, getting a quotient of 1 and remainder 488. Now we ask: what is the largest power of 2 that does not exceed 488? That's 256, or 2^8. Divide 488 by 256 and we get 232. (Note, at this point, that we are never going to get a quotient greater than 1, because then a higher power of 2 would be less than or equal to the number that we are dividing into.) The highest power of 2 that does not exceed is $128=2^7$. Applying the Division Algorithm again, we get $232=128+104$. Keep this up: we get $104=64+40$, $40=32+8$ and $8=8+0$. Putting everything together, we get: $1000=512+256+128+64+32+8$, or, putting it another way: $1000=2^9+2^8+2^7+2^6+2^5+2^3$. (All the "missing" powers of 2 appear with coefficient 0.) So, in base 2, we have $1000=(1111101000)_2$.

As another example, let us write 1000 in base 5. Reasoning similar to that used above gives $1000=625+375=625+3\ 125=5^4+3\ 5^3=(13000)_5$.

It is worth pointing out explicitly that since the base b can be any integer greater than 1, there is no reason why it can't be greater than 10. If we were to express, for example, 121 in base 12, we would start with $121=10\ 12^1+1\ 12^0$ and might be tempted to write $121=(101)_{12}$. But this is wrong, because $(101)_{12}=12^2+1=145$, not 121. So, instead of writing 10 or 11 as a coefficient, we need to invent a symbol for each of these—say, T and E. In that case, $121=(T1)_{12}$.

Exercises

1.44 Write the number 248 in base 2, base 3 and base 12.

Challenge Problems for Chapter 1

C1.1 Given ten consecutive integers, prove that one must be relatively prime to the other 9.

C1.2 Find all positive integer solutions x, y to the equation $x^y=y^x$.

C1.3 Prove that an integer $n>1$ has an odd number of divisors if and only if n is a square.

C1.4 If n is a positive integer, what are the possible values of $\gcd(n+1, n^2-n+1)$?

C1.5 If $p>3$ and $p+2$ are both primes, prove that 12 divides $2p+2$.

C1.6 If $n>1$ is an integer, prove that $1+\frac{1}{2}+\ldots+1/n$ is not.

C1.7 Let m and n be positive integers with gcd d. Prove that the number mn/d is a multiple of m and n, and divides any other multiple of m and n. We call this number the least common multiple of m and n, and denote it $[m, n]$.

C1.8   If a is a positive integer and $\sqrt{a}$ is rational, prove that a is a square. Using this, prove that if a and b are positive integers and $\sqrt{a}+\sqrt{b}$ is rational, then a and b are both squares.

C1.9   If n is a positive integer, let $d(n)$ be the number of positive divisors of n. Prove that there are infinitely many n for which $d(n) < d(n+1)$ and infinitely many for which $d(n) > d(n+1)$.

C1.10 Prove that the product of two consecutive positive integers is never a square.

2

Congruences and Modular Arithmetic

Let us begin with a simple motivating problem: If today is Monday, what day will it be 200 days from now? At first glance this seems like a cumbersome problem to answer, until we recognize a simple trick: 7 days from now, it will be Monday again. It will also be Monday 14 days from now, 21 days from now, etc. In particular, it will be Monday 196 days from now, because 196 is a multiple of 7. So, since we "reset the clock" at day 196, the original question posed is equivalent to asking what day it will be 4 days from now, and the answer to that is obvious: Friday.

The idea here (counting until we get to a certain number, then "equating" that number with 0, and resuming counting from scratch) is simple but very important in number theory. It is formalized in the definition of *congruence modulo n*, the concept to which we now turn.

2.1 Basic Definitions and Principles

Let us begin with a definition that captures the ideas spoken of above.

Definition 2.1.1

If a, b and n are integers, with n positive, we say that *a is congruent to b, modulo n* (written $a \equiv b \pmod{n}$), if n divides $a - b$.

So, for example, $200 \equiv 4 \pmod{7}$, which is the point of the motivating example above. Likewise, 10 is *not* congruent to 3 modulo 5. Question for the reader: why do we not bother defining "congruence modulo 0"?

The following theorem says, for those who know the terminology from a previous course, that "congruence modulo n" is an *equivalence relation*.

Theorem 2.1.2

Let a, b, c and n be integers, with n positive. Then

(i) $a \equiv a \pmod{n}$ (reflexivity)
(ii) if $a \equiv b \pmod{n}$, then $b \equiv a \pmod{n}$ (symmetry)
(iii) if $a \equiv b \pmod{n}$ and $b \equiv c \pmod{n}$, then $a \equiv c \pmod{n}$ (transitivity)

Proof. We prove (iii), and leave the similar (but easier) proofs of (i) and (ii) to the reader. To prove (iii), note that we are given that $n \mid a - b$ and $n \mid b - c$. It follows from our basic properties of divisibility that $n \mid (a - b) + (b - c)$ or, equivalently, $n \mid a - c$. But this is precisely the same as saying that $a \equiv c \pmod{n}$, which is what we wanted to prove.

There is a relationship between congruence and the Division Algorithm. Suppose we apply the Division Algorithm to the integers a and n, getting a quotient q and remainder r: $a = nq + r$. Then $a - r = nq$, so n divides $a - r$. Thus, every integer is congruent, mod n, to its remainder when divided by n. In particular, every integer is congruent to one of the integers $0, 1, \ldots, n - 1$. Moreover, no two different integers between 0 and $n - 1$ are congruent mod n (why?) so we have shown that every integer is congruent, mod n, to exactly one integer between 0 and $n - 1$. Another way to say this is to say that the set of integers $\{0, 1, \ldots, n - 1\}$ is a *complete set of residues mod n*. There are, of course, other complete sets of residues. If, for example, $n = 5$, then not only is $\{0, 1, 2, 3, 4\}$ a complete set of residues, but so is $\{10, 21, 67, 103, 14\}$.

Readers familiar with equivalence relations know that, given such a relation, we can define the *equivalence class* of an element of the set on which the relation is defined. In the case of congruence modulo n as the given equivalence relation, we call these *congruence classes*. The precise definition is as follows.

Definition 2.1.3

If n is a positive integer and a is any integer, then the *congruence class* determined by a, denoted $[a]_n$, is the set of all integers that are congruent to a modulo n.

If the integer n is understood, it is common to suppress the subscript in this notation and just write $[a]$. So, for example, if $n = 2$ then $[0]$ is the set of all even integers and $[1]$ is the set of all odd integers. (Make sure you understand why this is so.) If $n = 6$, then $[3] = \{ \ldots, -9, -3, 3, 9, 15, \ldots \}$. Our next theorem merely restates a familiar property about equivalence relations in general, but for those readers who are not familiar with this general theory, we state and prove the result from scratch.

Theorem 2.1.4

Any two distinct congruence classes modulo n are disjoint.

Proof. We will prove, equivalently, that if two congruence classes have even one element in common, then they are equal as sets. So, suppose [a] and [b] are two congruence classes modulo n that have the integer k in common. We will show [a]=[b] by showing that each of these sets is a subset of the other. To show that [a] ⊆ [b], let x be an arbitrary element of [a]. Then, by definition, $x \equiv a \pmod{n}$. But since k is another element in [a], we know that $a \equiv k \pmod{n}$. Hence, by transitivity, $x \equiv k \pmod{n}$. But k is also in [b], so $k \equiv b \pmod{n}$. By transitivity again, $x \equiv b \pmod{n}$, which is just another way of saying that x is an element of [b]. Hence, [a] ⊆ [b]. The proof of the reverse inclusion is identical (or we could just say that the result follows from symmetry, since a and b are interchangeable). This concludes the proof.

Since it is obvious from the definition that $a \in [a]$ (this is just the reflexive rule, rephrased), it follows from the previous theorem that [a]=[b] *if and only if* $a \equiv b \pmod{n}$. This is because if $a \equiv b \pmod{n}$, then $a \in [b]$; in that case, the congruence classes [a] and [b] both have the element a in common, and hence are equal. Conversely, if [a]=[b], then, since $a \in [a]$, it is also the case that $a \in [b]$, from which $a \equiv b \pmod{n}$ follows. This is a simple property of congruence classes, but an important one, and one that deserves to be stated explicitly.

Also, because $a \in [a]$, the union of the distinct congruence classes modulo n is all of $\mathbb{Z}$. Therefore, we can sum up some of the results of this section as follows: if n is a positive integer, and [a] denotes the congruence class of a modulo n, then any integer x lies in one, and only one, of the congruence classes [0], [1], ..., [n − 1].

We denote the set of these n congruence classes by $\mathbb{Z}_n$. Thus, for example, $\mathbb{Z}_2 = \{[0], [1]\}$, with [0] being the set of even integers and [1] the set of odd integers. In the next section of this book, we will define operations of addition and multiplication on the set $\mathbb{Z}_n$, thereby turning this set into a "miniature arithmetic system" (more precisely referred to, in algebraic terminology, as a "commutative ring with identity"; see Appendix C).

We end this section by discussing, in the spirit of the original motivating example, an easy but amusing application of congruences. Specifically, we want to prove that any year contains at least one Friday the 13th. We will do this under the assumption that our year has exactly 365 days (i.e., is not a leap year) but the same argument works, with minor arithmetic changes, for leap years as well. So, let us start with any non-leap year. January 13 of this year will fall on a certain day, let us number this day 0, and number the remaining days of the week, in order, 1, 2, ..., 6. (So, for example, if January 13 is a Wednesday, then Thursday is day 1, Friday is day 2, etc.) Now, what day is February 13? Because January has 31 days and 31 is congruent to 3 mod 7, February 13 is day 3. So is March 13, because March 13 occurs 28 days after February 13, and 28 is congruent to 0 mod 7. Proceeding in this way, we can determine what day of the week the 13th of each month falls on. After doing so, we see that all of the numbers from 0 to 6 appear as days of the week on which this happens. Since one of these numbers corresponds to Friday, it follows that there is a Friday the 13th in this year, and the result is proved.

Exercises

2.1 If a and b are integers, when is it true that $a \equiv b$ (mod 1)? When is it true that $a \equiv b$ (mod 2)?

2.2 If $a \equiv b$ (mod n), prove that $a^2 \equiv b^2$ (mod n). Can you generalize this?

2.3 Show by example that if x is a nonzero integer and $ax \equiv bx$ (mod n), it need not true that $a \equiv b$ (mod n).

2.4 In the previous exercise, show that if x and n are relatively prime, the conclusion $a \equiv b$ (mod n) *does* follow.

2.5 If a is any integer, show that $a^3 \equiv a$ (mod 3) and $a^5 \equiv a$ (mod 5). (Care to make a conjecture as to a more general result? Don't attempt to prove it, however.)

2.6 Prove that $4! \equiv -1$ (mod 5) and $6! \equiv -1$ (mod 7). (Again, care to make, without proof, a conjecture?)

2.2 Arithmetic in $\mathbb{Z}_n$

Fix a positive integer n, and let, as in the previous section, $\mathbb{Z}_n = \{[0], [1], \dots [n-1]\}$. We want to define operations of "addition" and "multiplication" in $\mathbb{Z}_n$, preferably in such a way as to maintain the "nice" properties of arithmetic (associative law, commutative law, etc.) that hold for the integers, or at least as many of these properties as we can. There is a pretty obvious way to approach this, namely by defining

$$[a] + [b] = [a + b] \tag{*}$$

but there is a fairly subtle potential problem that we must deal with in order for this definition to be "legal". The problem arises from the fact that the congruence class $[a]$ does not uniquely determine a.

To illustrate the problem, suppose a student, Joe, wants to compute $[2] + [7]$ in $\mathbb{Z}_8$. On the one hand, he can apply definition (*) above and get $[9]$, or $[1]$, since we want our answer to be an element of $\mathbb{Z}_8$. On the face of it, this is pretty straightforward. But now suppose that another student, Alice, also does this computation, but instead of writing $[2]$, she writes $[18]$. There's nothing wrong with this, because, in $\mathbb{Z}_8$, $[18]$ *is equal to* $[2]$. And suppose also that instead of writing $[7]$, she writes $[-1]$. Again, there's nothing wrong with this, because in $\mathbb{Z}_8$, [7] and [-1] *are exactly the same objects.* So, Alice, following the rule given in (*) above, will compute the sum to be [18]+ [-1]=[17]. Fortunately, [17] turns out to be the same object as [1], so Alice winds up with the same answer as Joe. If she hadn't, however, definition (*) would be worthless, because two people, both computing the sum correctly, would have arrived at different answers, based simply on the fact that they chose different representatives of the same object to compute with.

So, to define addition of congruence classes by the formula (*), we must show that what happened here is not a lucky coincidence and that it is *always* the case that this definition of addition is "independent of the representative" that we use to denote the congruence classes. In mathematics, this condition is expressed by saying that definition (*) is *well-defined*. Fortunately, it is quite simple to prove that addition is, indeed, well-defined; this is the content of our next theorem.

Theorem 2.2.1

The formula $[a]+[b]=[a+b]$ for addition in $\mathbb{Z}^n$ is well-defined.

Proof. What we must show is that, in $\mathbb{Z}_n$, if $[a]=[c]$ and $[b]=[d]$, then $[a+b]=[c+d]$. Translated into the language of congruences, our assumptions are that $a \equiv c$ (mod n) and $b \equiv d$ (mod n), which in turn means that $n \mid a - c$ and $n \mid b - d$. But it then follows from our basic properties of divisibility that $n \mid (a - c)+(b - d)$, or $n \mid (a + b) - (c + d)$, which is exactly what we needed to show.

This result can be rephrased as saying: "congruences can be added". Specifically, if $a \equiv c$ (mod n) and $b \equiv d$ (mod n), then $a+b \equiv c+d$ (mod n).

Now that we know that addition in $\mathbb{Z}_n$ is well-defined, it is very easy to prove some basic properties about this operation. The following theorem summarizes some of them. The proof of this result is left as an exercise, since it simply amounts to mechanically using the definition, keeping in mind that these properties are already known to be true in $\mathbb{Z}$.

Theorem 2.2.2

Addition in $\mathbb{Z}_n$ satisfies the following properties for all integers a, b and c:

(a) Associativity: $[a] +([b]+[c])=([a]+[b])+ [c]$
(b) Additive identity: $[a]+[0]=[a]$
(c) Additive inverses: $[a]+[-a]=[0]$
(d) Commutative law: $[a]+[b]=[b]+[a]$

As pointed out in Appendix C, this theorem says that $\mathbb{Z}_n$, with respect to the operation of addition, is an *abelian group*.

The additive inverse of an element $[a]$ in $\mathbb{Z}_n$ is denoted $- [a]$. This allows us to define subtraction in $\mathbb{Z}_n$: $[a] - [b]$ is simply defined to be sum of $[a]$ and $- [b]$.

We caution the reader, however, not to think of $-[a]$ as a "negative number", for the simple reason that the concepts of "positive" and "negative" make no sense in $\mathbb{Z}_n$. For example, in $\mathbb{Z}_{12}$, $-[11]=[-11]=[1]$. Certainly [1] cannot be both positive and negative! The minus sign appearing before an element of $\mathbb{Z}_n$ simply means "additive inverse of".

We next consider an operation of *multiplication* on $\mathbb{Z}_n$. As was the case with addition, the definition of multiplication is an easy and natural one, but we need to prove that it is well-defined. We simply define

$$[a][b] = [ab] \qquad (**)$$

Theorem 2.2.3

The formula $[a][b]=[ab]$ for multiplication in $\mathbb{Z}_n$ is well-defined.

Proof. As was the case with addition, what we must show is that, in $\mathbb{Z}_n$, if $[a]=[c]$ and $[b]=[d]$, then $[ab]=[cd]$. Translated into the language of congruences, our assumptions are that $a \equiv c \pmod{n}$ and $b \equiv d \pmod{n}$, which in turn means that $n \mid a-c$ and $n \mid b-d$. We want to show that $n \mid ab-cd$. To do this, we note that $ab-cd=ab-ad+ad-cd=a(b-d)+(a-c)d$. And now we're done, because we know that $(b-d)$ and $(a-c)$ are both divisible by n, and hence so is $a(b-d)+(a-c)d$.

As was the case with addition, this result can be rephrased as saying "congruences can be multiplied": if $a \equiv c \pmod{n}$ and $b \equiv d \pmod{n}$, then $ab \equiv cd \pmod{n}$.

This fact lies at the heart of an interesting arithmetic result that is sometimes taught to children—namely, that a multi-digit number is divisible by 9 if and only if the sum of its digits is. To see why this is so, begin by noting the obvious congruence $10 \equiv 1 \pmod{9}$. Multiplying this congruence by itself repeatedly yields $10^k \equiv 1 \pmod{9}$ for any positive integer k. Now take a multi-digit number, say 341. This can be written as
$3 \cdot 10^2 + 4 \cdot 10^1 + 1 \cdot 10^0$, which, given the observation we just made, is congruent mod 9 to $3+4+1$, the sum of its digits. So, the number itself and the sum of its digits are congruent mod 9, which means that the number is congruent to 0 mod 9 if and only if the sum of its digits is. A similar test works for divisibility by 3, because $10 \equiv 1 \pmod{3}$. Can you formulate a divisibility test for divisibility by 11?

Like addition, multiplication in $\mathbb{Z}_n$ inherits certain basic arithmetic properties from the set of integers. They are collected in the next theorem, whose proof, like the previous one, is left to the reader as an easy exercise in "unrolling" the definition of multiplication.

Theorem 2.2.4

Multiplication in $\mathbb{Z}_n$ satisfies the following properties for all integers a, b and c:

(a) Associativity: $[a]([b][c])=([a][b])[c]$
(b) Multiplicative identity: $[a][1]=[a]$

(c) Commutative law: $[a][b] = [b][a]$

(d) Distributive law: $[a]([b] + [c]) = [a][b] + [a][c]$

In the language of Appendix C, the properties set out in this theorem and Theorem 2.2.2 say that $\mathbb{Z}_n$ is, with respect to the operations of addition and multiplication, a *commutative ring* (or, depending on which book you read, a *commutative ring with identity*).

We thus have created a miniature "arithmetic system" that shares some properties with the ring of integers, but which in some other ways is quite different from this ring. This new arithmetic system, for example, is, unlike the set of integers, a finite set. The operations of addition and multiplication also can act quite differently than the corresponding operations on the set of integers. For example, note that in the set of integers, it is not possible to multiply two nonzero integers and get 0. However, in the set $\mathbb{Z}_n$, it is, for certain integers n, quite possible to multiply two nonzero elements in $\mathbb{Z}_n$ ("zero" in the set $\mathbb{Z}_n$ refers to the additive identity $[0]$) and get $[0]$. For example, in $\mathbb{Z}_8$, we have $[2][4] = [0]$. The elements $[2]$ and $[4]$ are called *zero divisors* in $\mathbb{Z}_n$. Our next theorem states exactly when zero divisors can exist in $\mathbb{Z}_n$.

Theorem 2.2.5

Let $n > 1$. The ring $\mathbb{Z}_n$ has no (nonzero) zero divisors if and only if n is prime.

Proof. Suppose first that n is prime and that $[a][b] = [0]$. We want to show that either $[a] = [0]$ or $[b] = [0]$. This, however, follows immediately from Euclid's Lemma (Theorem 1.5.2): if $[ab] = [0]$ in $\mathbb{Z}_n$, that means n divides ab, but since n is prime, Euclid's Lemma says that n divides a or n divides b. This in turn means either $[a] = [0]$ or $[b] = [0]$, as was to be proved.

For the converse, suppose that n is not a prime. Then we can write $n = ab$, where both a and b are strictly between 1 and n. But this means that $[0] = [n] = [a][b]$, where $[a]$ and $[b]$ are nonzero elements in $\mathbb{Z}_n$; i.e., $\mathbb{Z}_n$ has zero divisors.

In algebraic language, this theorem says that $\mathbb{Z}_n$ (like $\mathbb{Z}$) is an *integral domain* if and only if n is prime.

Let us compare and contrast $\mathbb{Z}_n$ and $\mathbb{Z}$ in another respect. Suppose we look for the nonzero integers a in $\mathbb{Z}$ that have multiplicative inverses—i.e., an integer b such that $ab = 1$. Clearly, the only integers a with this property are ± 1. The situation, however, is different in $\mathbb{Z}_n$, where $[1]$, the multiplicative identity, plays the role that 1 plays in $\mathbb{Z}$. Consider, for example, $\mathbb{Z}_8$, where we can quickly see that $[1]$, $[3]$, $[5]$ and $[7]$ all have multiplicative inverses: $[3][5] = [1]$, and $[7][7] = [1]$. We can equally quickly see, by simply considering all possible products, that $[2]$, $[4]$ and $[6]$ do not have multiplicative inverses. See a pattern? The next theorem spells it out.

Theorem 2.2.6

The nonzero element $[a]$ in $\mathbb{Z}_n$ has a multiplicative inverse if and only if a and n are relatively prime.

Proof. First, suppose that $[a]$ has a multiplicative inverse $[b]$. Then $[ab]=[1]$, which means (translating into the language of congruences) that $ab \equiv 1 \pmod{n}$. This means that n divides $ab - 1$, or, by definition, that $ab - 1 = nx$ for some integer x. But this means that $ab - nx = 1$. Since we can express 1 as a linear combination of a and n, this means that a and n are relatively prime. For the converse, simply reverse this line of reasoning.

We use the term *unit* to refer to a nonzero element of $\mathbb{Z}_n$ that has a multiplicative inverse. Note that if n is a prime, then every integer a between 1 and $n - 1$ is relatively prime to n, and so *every* nonzero element of $\mathbb{Z}_n$ is a unit. In algebraic terminology (see Appendix C once again), $\mathbb{Z}_n$ is a *field*. The ring of integers, however, is not a field.

We will denote the set of all units in $\mathbb{Z}_n$ by $\mathbb{Z}_n^*$. Thus, for example, $\mathbb{Z}_8^* = \{[1], [3], [5], [7]\}$. Note that this set is not closed under addition but is closed under multiplication. This is true in general (why?). $\mathbb{Z}_n^*$, with respect to the operation of multiplication, is an algebraic structure known as a *group* (see Appendix C). Of course, when n is a prime, $\mathbb{Z}_n^*$ is simply the set of *all* nonzero congruence classes mod n.

We end this section by briefly addressing two computational questions: first, if $[a]$ is a unit in $\mathbb{Z}_n$, how do we explicitly find its multiplicative inverse? Since our assumption is that a and n are relatively prime, we know there are integers x and y such that $ax + ny = 1$; we also know that we can use the Euclidean Algorithm to explicitly find x and y. But then (consider this equation modulo n) it follows immediately that $[x]$ is a multiplicative inverse of $[a]$. It is also the only one (see Exercise 2.10).

Second, there is a technique ("repeated squaring") that allows for reasonably efficient exponentiation modulo a given number. Recall that any positive integer can be written in "base 2" as a sum of powers of 2. For example, $37 = 32 + 4 + 1 = 2^5 + 2^2 + 2^0$. Suppose we wanted to compute 5^{37} modulo 17. Computing the 37th power of a number is no fun, but we can avoid that by using congruences. Begin by repeatedly squaring 5 and reducing mod 17:

$$5^2 \equiv 8$$

$$5^4 \equiv 13 \equiv -4$$

$$5^8 \equiv 16 \equiv -1$$

$$5^{16} \equiv 1$$

$$5^{32} \equiv 1 \text{ (all congruences, of course, being mod 17)}$$

Thus, $5^{37} = 5^{32}5^4 5 \equiv (13)(8) \equiv 2$. This method, of course, generalizes.

Exercises

2.7 Prove Theorems 2.2.2 and 2.2.4.

2.8 Determine (without electronic assistance) the remainder when 3^{85} is divided by 11.

2.9 Find every element in $\mathbb{Z}_{12}$ that has a multiplicative inverse and, for each such element, find that inverse. Then show that every non-zero element that does not have a multiplicative inverse is a zero divisor.

2.10 If a and n are relatively prime, prove that the multiplicative inverse of $[a]$ in $\mathbb{Z}_n$ is unique.

2.11 Students sometimes wonder why we don't define addition of fractions by the rule $(a/b)+(c/d)=(a+c)/(b+d)$. Show that this "definition" is not well-defined.

2.3 Linear Equations in $\mathbb{Z}_n$

Back in the lower grades, students, after learning about the integers and rational numbers, learn how to solve equations in these systems. Now that we have the new arithmetic system $\mathbb{Z}_n$ to play with, it is natural to look at equations in this system. The simplest equations are the linear ones, so we will start with these.

So, consider

$$[a][x] = [b]$$

in $\mathbb{Z}_n$. This is equivalent to the statement

$$ax \equiv b \pmod{n}$$

which in turn means the same thing as

$$ax - ny = b$$

for some integer y. We know from Section 1.3 that this equation has a solution in integers x and y if and only if the greatest common divisor of a and n divides b. We thus have the following result, which really is just a restatement of previously established results in new language: *the equation $[a][x]=[b]$ in $\mathbb{Z}_n$ has a solution if and only if the greatest common divisor of a and n divides b.*

In fact, we can say more. In Section 1.3, we learned not only when a solution to $ax - ny=b$ exists, but also what the general form of a solution looks like. We can translate that language to the language of linear equations in $\mathbb{Z}_n$ as well. The result is the next theorem.

Theorem 2.3.1

The congruence equation $ax \equiv b \pmod{n}$ has an integer solution if and only if d, the greatest common divisor of a and n, divides b. If this condition is satisfied, then there are exactly d incongruent solutions modulo n. If X is a particular solution, then any other solution to this equation can be written in the form $X + t(n/d)$ for some integer t.

Proof. We have already proved the first sentence of this theorem. Now suppose that d divides b and that X satisfies $aX \equiv b \pmod{n}$. Let x' be any other solution to this equation. We will mimic the proof of Theorem 1.3.10 to prove that x' has the desired form. It follows from the assumption that $aX \equiv ax'$ $\pmod{n}$. This means that n/d divides $(a/d)(x' - X)$, from which it follows, as in Theorem 1.3.10, that n/d divides $(x' - X)$. This means that

$(x' - X) = t(n/d)$ for some integer t, thus proving the first part of the theorem.

It remains to be shown that there are exactly d incongruent solutions modulo n. In fact, choosing $t = 0, 1, \ldots, d - 1$ yields these incongruent solutions modulo n. This is easy to see, and we leave the details as an exercise.

Observe that Theorem 2.2.6 and Exercise 2.10 are both special cases of this theorem. When a and n are relatively prime, we can solve the equation $ax \equiv b$ $\pmod{n}$ by simply multiplying by the multiplicative inverse, mod n, of a. As an example, consider the equation

$3x \equiv 7 \pmod{11}$. Since 3 and 11 are relatively prime, we know that [3] is a unit in $\mathbb{Z}_{11}$ or, what amounts to the same thing, there is an integer x such that $3x \equiv 1 \pmod{11}$. Mental arithmetic (or the Euclidean Algorithm) tells us that 4 works. Multiplying both sides of $3x \equiv 7 \pmod{11}$ by 4 (remember, we can multiply congruences) and keeping in mind that 28 is congruent to 6 mod 11 gives 6 as the solution to the congruence.

We next turn our attention to *systems* of congruence equations and look for integers that satisfy all of them simultaneously. To illustrate the basic ideas, let's start with a simple example of two equations:

$$x \equiv 4 \pmod{5}$$

$$x \equiv 7 \pmod{11}$$

We know that 4 is the unique solution mod 5 to the first equation, but we also have to consider things mod 11 to satisfy the second. We also know (by the previous theorem) that the general solution to the first equation is $4 + 5t$ for some integer t. So, if we want an x that satisfies both equations, the sensible thing to do is plug $4+5t$ into the second, thus getting

$$4 + 5t \equiv 7 \;(\text{mod } 11) \quad \text{or}$$

$$5t \equiv 3 \;(\text{mod } 11).$$

As we just observed, however, we know how to solve this last equation: find a multiplicative inverse of 5 mod 11. A few minutes of thought gives the answer: $x=9$. Multiplying the equation $5t \equiv 3 \;(\text{mod } 11)$ by 9 gives $t=5$, so $x=4+5t=29$. Observe that 29 does, indeed, satisfy both of the two original equations above. Now, if there were some other solution, say y, to both of these equations, then we would have $x \equiv y \;(\text{mod } 5)$ and $x \equiv y \;(\text{mod } 11)$; since 5 and 11 are relatively prime and both divide $x - y$, this would mean that $x - y$ is also divisible by 55 (see Theorem 1.3.10); i.e., that $x \equiv y \;(\text{mod } 55)$. So we have not only shown that a solution exists but also shown that it is unique modulo 55.

The method of reasoning employed above generalizes and allows a simple proof of the following theorem, which is a special case of a result called the Chinese Remainder Theorem.

Theorem 2.3.2

If m and n are relatively prime positive integers, and a and b are any two integers, then the two congruence equations $x \equiv a \;(\text{mod } m)$ and $x \equiv b \;(\text{mod } n)$ have a simultaneous solution, and any two solutions are congruent mod mn.

Proof. As above, consider any integer of the form $a+mt$, where t is an integer. Any integer of this form satisfies $x \equiv a \;(\text{mod } m)$; we want to show that there is an integer t such that this integer also satisfies $x \equiv b \;(\text{mod } n)$. In other words, we ask: can we find a t such that $a+mt \equiv b \;(\text{mod } n)$? This amounts to asking whether there is a t for which $mt \equiv b - a \;(\text{mod } n)$, and we know the answer to this question is "yes" because m, being relatively prime to n, has a multiplicative inverse mod n, so we can solve for t by multiplying by this multiplicative inverse. Once we find t, then $a+mt$ is a simultaneous solution to the system of congruence equations. The uniqueness of this solution mod mn follows from the fact that the difference between any two solutions is divisible by m and by n, and hence (because m and n are relatively prime) by mn, just as in the example above.

We can now state and prove the more general version of the Chinese Remainder Theorem, which involves a system of k equations rather than just two. We could actually use Theorem 2.3.2 to prove the more general case, but it seems advisable to give a more constructive proof.

Theorem 2.3.3

Suppose that $n_1, ..., n_k$ are k positive integers, any two of which are relatively prime. Suppose also that $a_1, ..., a_k$ are any k integers. Then there is an integer x that simultaneously satisfies each of the following congruences:

$$x \equiv a_1 \,(\mathrm{mod}\, n_1)$$
$$\vdots$$
$$x \equiv a_k \,(\mathrm{mod}\, n_k).$$

Moreover, any two such solutions are congruent mod $N = n_1, ..., n_k$.

Proof. We will prove the last sentence first. If x and y are two simultaneous solutions to this system of congruences, then, by the basic equivalence relation properties of congruence relation, it must be the case that $x \equiv y \pmod{n_i}$ for reach i between 1 and k. But because the n_i are relatively prime, it must be the case that $x \equiv y \pmod{N}$.

We now prove that such an x exists. First, let $N_i = N/n_i$. In other words, N_i is simply the product of all the n's except for n_i. Observe that N_i and n_i are relatively prime: if some prime p divided N_i, then by Euclid's Lemma p would have to divide some n_j ($j \neq i$) and hence p could not, by our assumption of pairwise relative primeness, also divide n_i.

Because N_i and n_i are relatively prime, N_i has a multiplicative inverse x_i mod n_i: i.e., $x_i N_i \equiv 1 \pmod{n_1}$. Now let $x = a_1 x_1 N_1 + \backslash ... + a_k x_k N_k$. If i is any integer between 1 and k, then every summand of x except $a_i x_i N_i$ is congruent to 0 mod n_i (why?) and this one summand is congruent to a_i mod n_i. Hence, x is congruent to a_i mod n_i and is therefore a simultaneous solution to the system of congruences.

We close this section with a bit of history: the term "Chinese Remainder Theorem" memorializes an ancient Chinese document, likely dating back to the late 3rd century, called the *Mathematical Manual*. In this document, Sun Tze poses the problem of finding an integer that leaves a remainder of 2 when divided by 3, a remainder of 3 when divided by 5 and a remainder of 2 when divided by 7. Sun Tze provides an answer; you will be offered the opportunity to provide one in the exercises that follow.

Exercises

2.12 Find the smallest positive integer that is simultaneously congruent to 1 mod 2, congruent to 2 mod 3 and congruent to 3 mod 5. Then, having found the smallest positive integer that satisfies these congruences, find the next-smallest one.

2.13 Compute the product of all nonzero elements of $\mathbb{Z}_5$. Express your answer as an element of $\mathbb{Z}_5$. Then do the same for $a \equiv b_7$. Care to guess what the answer would be if we did the same calculation for $\mathbb{Z}_{101}$?

2.14 Find a solution to the congruence $3x \equiv 2 \pmod{17}$.

2.15 Show that Theorem 2.3.2 is not true if m and n are not assumed to be relatively prime.

2.16 If n is any positive integer, prove that there are n consecutive integers, each one divisible by the square of a prime.

2.17 Solve Sun Tze's problem, quoted above.

2.18 Prove that, if p is an odd prime, there are two consecutive integers whose sum is divisible by p.

2.4 The Euler Phi Function

We have seen in the previous section that there is some value to knowing, for a positive integer n, whether an integer a is or is not relatively prime to n. We will shortly see that it is also useful to know just how many integers there are, between 1 and n, with this property.

Definition 2.4.1

If n is a positive integer, then the number of integers between 1 and n, inclusive, that are relatively prime to n, is denoted $\varphi(n)$, where φ is the Greek letter phi. This defines a function φ from the set of positive integers into itself, called the Euler phi function.

We say "between 1 and n" rather than "between 1 and $n-1$" to handle the case $n=1$; by this definition, $\varphi(1)=1$. Obviously, if n is greater than 1, we can just say "between 1 and $n-1$" because n is never going to be relatively prime to itself. For another example, $\varphi(10) = 4$ because there are four positive integers between 1 and 10 that are relatively prime to 10, namely 1, 3, 7 and 9. Likewise, $\varphi(15)=8$; check this yourself. Also, $\varphi(101)=100$; we can see this without having to do any tedious calculations by just noticing that since 101 is a prime, *all* the integers from 1 to 100 are relatively prime to 101. This last observation, of course, can be generalized:

Theorem 2.4.2

If n is a positive integer that is greater than 1, then n is prime if and only if $\varphi(n)=n-1$.

Proof. If n is a prime, then, by the reasoning above, $\varphi(n)=n-1$. Conversely, if n is not a prime, then n has a nontrivial divisor, say a. In this case, of course, a is not relatively prime to n. So, among the integers from 1 to $n-1$, there is at least one that is not relatively prime to n, so $\varphi(n) < n-1$.

Now that we know how φ treats primes, it is only natural to ask how φ treats prime powers. There's a simple answer.

Theorem 2.4.3

If p is a prime and k is a positive integer, then $\varphi(p^k)=p^k - p^{k-1}$.

Proof. Of all the integers from 1 to p^k, the only ones that are *not* relatively prime to p^k are those that are multiples of p. There are p^{k-1} of them: p, $2p$, ..., $p^{k-1}p$.

There is another way to think about $\varphi(n)$, at least when n is greater than 1. Recall from Section 2.2 that an element $[a]$ in $\mathbb{Z}_n$ is a unit (i.e., has a multiplicative inverse) if and only if a and n are relatively prime. Thus, $\varphi(n)$ is equal to the number of units in $\mathbb{Z}_n$.

Our next theorem tells us how φ treats products, or at least products of relatively prime integers. In combination with Theorem 2.4.3, this theorem will allow us to establish a formula for φ. Because the proof of the theorem is not trivial, we will defer it to the end of the section.

Theorem 2.4.4

If m and n are relatively prime positive integers, then $\varphi(mn)=\varphi(m)\varphi(n)$.

The first thing to note is that this theorem admits an easily proved (by mathematical induction) generalization. We state it and leave the proof to the reader.

Theorem 2.4.5

If n_1, ..., n_k are k positive integers, any two of which are relatively prime, then $\varphi(n_1 \ldots n_k)=\varphi(n_1) \ldots \varphi(n_k)$.

If the prime factorization of an integer n is known, this result allows an easy calculation of $\varphi(n)$.

Theorem 2.4.6

If $n=\prod p_i^{k_i}$ is the prime factorization of the integer $n>1$, then $\varphi(n)=\prod p_i^{k_1}\left(1 - 1/p_i\right)=n \prod(1 - 1/p_i)$.

Proof. The first equality follows immediately from applying Theorem 2.4.5 to the product of prime powers defining n, and then applying Theorem 2.4.3 to each prime power in that product. The second equality follows from collecting terms, using the fact that $n=\prod p_i^{k_i}$.

Of course, it still remains to prove Theorem 2.4.4. We tie up this loose end now. We can assume both m and n are greater than 1, as otherwise the result is trivial. We will prove the theorem by counting units in the sets $\mathbb{Z}_n$, $\mathbb{Z}_m$ and $\mathbb{Z}_{mn}$. (To help keep things straight, we will use subscripts to keep track of congruence classes relative to different moduli.) Recall that if G and H are sets, then G×H, the *Cartesian product* of G and H, denotes the set of all ordered pairs whose first component comes from G and whose second component comes from H. We assume the reader knows that if G has m elements and H has n elements, then G×H has mn elements. In particular, $\mathbb{Z}_m \times \mathbb{Z}_n$ has mn elements, the same number of elements as the set $\mathbb{Z}_{mn}$.

Define a function T from $\mathbb{Z}_{mn}$ to the Cartesian product $\mathbb{Z}_m \times \mathbb{Z}_n$ as follows: if $[a]_{mn}$ denotes an arbitrary element of $\mathbb{Z}_{mn}$, let $T([a]_{mn}) = ([a]_m, [a]_n)$. First observe that this function is clearly well-defined (see the definition in Section 2.2); this follows from the fact that two integers congruent mod mn are also congruent mod m and mod n. Moreover, this function is onto: if we start with any two residue classes $[a]_m$ and $[b]_n$ in $\mathbb{Z}_m$ and $\mathbb{Z}_n$, respectively, then by the Chinese Remainder Theorem there is an integer x that is congruent to a mod m and to b mod n; it follows that $T([x]_{mn}) = ([a]_m, [b]_n)$. Since a function from a set onto another set of the same size must also be 1-1, it follows that T is a bijection from $\mathbb{Z}_{mn}$ to $\mathbb{Z}_m \times \mathbb{Z}_n$.

What we are really interested in, however, is the way T acts on the subset of units of $\mathbb{Z}_{mn}$. Let us denote this set as $\mathbb{Z}_{mn}^*$ and use similar notation for $\mathbb{Z}_m$ and $\mathbb{Z}_n$. We claim that T is a bijection from $\mathbb{Z}_{mn}^*$ to the set $\mathbb{Z}_m^* \times \mathbb{Z}_n^*$. Since the set $\mathbb{Z}_{mn}^*$ has $\varphi(mn)$ elements and the set $\mathbb{Z}_m^* \times \mathbb{Z}_n^*$ has $\varphi(m)\,\varphi(n)$ elements, this will prove the result.

We first observe that T actually maps $\mathbb{Z}_{mn}^*$ into the set $\mathbb{Z}_m^* \times \mathbb{Z}_n^*$. This is pretty clear: if $[a]_{mn}$ is an element of $\mathbb{Z}_{mn}^*$, then a is relatively prime to mn. But then a is also relatively prime to both m and n, so $[a]_m$ is an element of $\mathbb{Z}_m^*$ and $[a]_n$ is an element of $\mathbb{Z}_n^*$.

We next show that T is 1-1. Of course, this follows from the bijectivity of T as a function from $\mathbb{Z}_{mn}$ to $\mathbb{Z}_m \times \mathbb{Z}_n$ (why?) but let us give a simple direct proof. Suppose $T([a]_{mn}) = T([b]_{mn})$. Then by definition of T, we must have $[a]_m = [b]_m$ and $[a]_n = [b]_n$. But then, since m and n are relatively prime, it follows that $[a]_{mn} = [b]_{mn}$, as was to be proved.

Finally, we show that T, as a function from $\mathbb{Z}_{mn}^*$ into the set $\mathbb{Z}_m^* \times \mathbb{Z}_n^*$, is onto. Let us start with arbitrary residue classes $[a]_m$ and $[b]_n$ in $\mathbb{Z}_m^*$ and $\mathbb{Z}_n^*$. We know (see above) that there is an integer x such that $T([x]_{mn}) = ([a]_m, [b]_n)$. We know $[x]_{mn}$ is an element of $\mathbb{Z}_{mn}$, but is it an element of $\mathbb{Z}_{mn}^*$? In fact, it is: we know that x is congruent to a mod m, and since a is relatively prime to m, this means that x must be also. The same reasoning shows that x is relatively to n. But if x is relatively prime to both m and n, and m and n are relatively prime, then x must be relatively prime to mn, as needed to be shown.

We have shown that T induces a bijection from the set $\mathbb{Z}_{mn}^*$ onto the set $\mathbb{Z}_m^* \times \mathbb{Z}_n^*$. The desired result then follows immediately.

Exercises

2.19 Compute $\varphi(n)$ for every positive integer n between 1 and 25, inclusive.

2.20 Based on your calculations above, formulate and prove a conjecture about the parity of $\varphi(n)$.

2.21 Compute $\varphi(1076)$ and $\varphi(888)$.

2.22 Find, with proof, all n for which $\varphi(n)=n-2$.

2.23 Prove that $\varphi(2n)=\varphi(n)$ if and only if n is odd.

2.24 Find, with proof, all n for which $\varphi(3n)=3\varphi(n)$.

2.25 Prove that, for every positive integer k, there are only a finite number of n for which $\varphi(n)=k$. (There may not be any at all.)

2.26 Prove that if n is a positive integer, then $\varphi(n2)=n\ \varphi(n)$.

2.27 If n is greater than 1, prove that the sum of all integers between 1 and $n-1$ that are relatively prime to n is $n\ \varphi(n)/2$.

2.28 Find, with proof, all positive integers n for which $\varphi(n)=6$.

2.29 Prove that there are no positive integers n for which $\varphi(n)=14$.

2.5 Theorems of Wilson, Fermat and Euler

In this section of the text, we explore three classical theorems related to congruences; the first two involve prime moduli and the third extends the second in cases where the modulus is not necessarily prime. To state the first result, known as *Wilson's Theorem*, we need the notion of the factorial of a number.

Definition 2.5.1

If n is a positive integer, we define n factorial (denoted $n!$) to be the product of all integers from 1 to n, inclusive.

So, for example, $3!=6$ and $5!=120$. It is also customary to define $0!=1$.

Before stating and proving Wilson's Theorem, we need a simple result concerning elements of $\mathbb{Z}_p$ that are their own multiplicative inverses.

Theorem 2.5.2

If p is a prime, and $[a]$ is an element of $\mathbb{Z}_p$ that is its own multiplicative inverse, then $[a]=[1]$ or $[a]=[p-1]$.

Proof. We are told that $[a^2] = [1]$, which means (translated into the language of congruences) that $a^2 \equiv 1 \pmod{p}$. This in turn means that p divides
$a^2 - 1 = (a - 1)(a + 1)$, and by Euclid's Lemma, this means that either p divides $(a - 1)$ or p divides $(a + 1)$. In the first case, $[a] = [1]$ and in the second case $[a] = [-1] = [p - 1]$.

We can now state and prove Wilson's Theorem. The proof exploits the multiplicative structure of $\mathbb{Z}_p$.

Theorem 2.5.3

(Wilson) If p is a prime, then $(p - 1)! \equiv -1 \pmod{p}$.

Proof. We can assume $p > 2$, as the result is clearly true when $p = 2$. We will prove the equivalent statement that, in $\mathbb{Z}_p$, $[(p - 1)!] = [-1]$. Note that $[(p - 1)!]$ is just the product of all nonzero elements of $\mathbb{Z}_p$. Because multiplication is commutative, we can rearrange the terms in this product, and thus write $[(p - 1)!] = [1][p - 1]X$, where "X" denotes the product of the remaining terms. But by Theorem 2.4.2, no term in the product defining X is its own inverse, so when writing out this product, we can pair off every term with its (different) multiplicative inverse. It follows that X just consists of a product of [1] terms and is therefore equal to [1]. Thus $[(p - 1)!] = [1]$ $[p - 1]X = [p - 1] = [-1]$, as was to be proved.

Our next theorem is often referred to as Fermat's Little Theorem (FLT). The word "little" in the title is not intended to denigrate the result, which is important and interesting, but to distinguish this result from the much more famous result (see Chapter 0) that bears Fermat's name.

Theorem 2.5.4

(Fermat) If p is a prime, and a is an integer that is relatively prime to p, then $a^{p-1} \equiv 1 \pmod{p}$.

Proof. Begin by listing the nonzero elements of $\mathbb{Z}_p$:

$$[1], [2], \ldots, [p-1] \tag{*}$$

and now multiply each of them by $[a]$:

$$[1a], [2a], \ldots, [(p-1)a] \tag{**}$$

Observe that there is no duplication among the elements enumerated in (**): if, for example, we had $[ax] = [ay]$, then multiplying both sides by the multiplicative inverse of $[a]$ would give $[x] = [y]$. Thus, since the elements comprising (*) are all distinct, so are the elements comprising (**). But there are $p - 1$ of them, so they must consist of all the nonzero elements of $\mathbb{Z}_p$, albeit perhaps in a different order. But this means that the *product* of all the terms listed in (*) is the same as the *product* of the terms listed in (**):

$$[(p-1)!] = [a^{(p-1)}][(p-1)!]$$

from which it immediately follows, by Wilson's Theorem (or simple cancellation), that

$$[1] = [a^{(p-1)}]$$

which is clearly equivalent to the result that we wanted to prove.

There is another result, closely related to this theorem, that also goes by the name Fermat's Little Theorem. It discusses what happens if we remove the assumption that a is an integer that is relatively prime to p.

Theorem 2.5.5

(Fermat). If p is a prime, and a is any integer, then $a^p \equiv a \pmod{p}$.

Proof. If a is relatively prime to p, then Theorem 2.4.4 holds; multiply both sides of $a^{p-1} \equiv 1 \pmod{p}$ by a to obtain the desired result. If a is not relatively prime to p, then it must be a multiple of p, and the theorem again follows because both a^p and a are congruent to 0 mod p, and hence are congruent to each other.

Fermat's Theorem has several uses. One is to simplify (in certain cases) exponentiation modulo a prime p. We have seen in Section 2.2 that this can be done via successive squaring; FLT provides an alternative approach. We illustrate with an example: let us determine, for example, what the remainder is when 3^{82} is divided by 17. By FLT (since 17 is a prime), $3^{16} \equiv 1 \pmod{17}$. Raising both sides of this congruence to the 5th power yields $3^{80} \equiv 1 \pmod{17}$. It follows that $3^{82} = 3^{80}3^2$ is congruent to 9 mod 17, so this is the remainder.

FLT can also be used as a means of determining that a number is not prime. Suppose a very large integer p is given and it is desired to know whether it is prime or not. Even with modern computers, factoring a number is computationally infeasible; however, as we have seen, exponentiation modulo a prime is more tractable. If we can find an integer a that is relatively prime to p (this can be determined, remember, by the Euclidean Algorithm) and for which a^{p-1} is not congruent to 1 mod p, then it must be the case that p is not prime.

Unfortunately, this is not an "if and only if" result. Composite integers n exist with the property that $a^{n-1} \equiv 1 \pmod{n}$ for all a relatively prime to n. An example is $561 = 3 \times 11 \times 17$. To see why this is so, suppose that a is relatively prime to 561. To show that $a^{560} \equiv 1 \pmod{561}$, note first that it suffices to show this congruence holds mod 3, 11 and 17 (why?). Now, by FLT, since a is clearly relatively prime to 3, we have $a^2 \equiv 1 \pmod{3}$. Raising both sides to the 280th power gives the desired congruence. Likewise, another application of FLT gives $a^{10} \equiv 1$ (mod 11), from which we immediately conclude that $a^{560} \equiv 1$ (mod 11). The same reasoning applies for the prime 17 as well, concluding the proof.

A composite integer n, greater than 1, with the property that $a^{n-1} \equiv 1$ (mod n) for all a relatively prime to n, is called a *Carmichael number*. We have just shown that 561 is one; there are others. In fact, it was proved in 1994 by Alford, Granville and Pomerance that there are infinitely many.

FLT clearly does not hold if the modulus is not a prime: 2^3 is certainly not congruent to 1 mod 4, for example. But there is a generalization of FLT, called Euler's Theorem, which does hold; it makes use of the Euler phi function from the last section. As you read the statement of the theorem, note that if n is a prime, it reduces to FLT.

Theorem 2.5.6

(Euler) If n is a positive integer, and a is relatively prime to n, then $a^{\varphi(n)} \equiv 1$ (mod n).

Proof. The proof of this is almost identical to the proof of FLT 2.4.4. As before, let us denote by $\mathbb{Z}_n{}^*$ the set of units of $\mathbb{Z}_n$—i.e., the set of elements of this set that have multiplicative inverses. There are $\varphi(n)$ such elements, and the set $\mathbb{Z}_n{}^*$ is closed under multiplication. If we list the $\varphi(n)$ elements of $\mathbb{Z}_n{}^*$ (starting, say, with [1] and ending with [$p-1$]} and then multiply each one by [a], we again get $\varphi(n)$ distinct elements of $\mathbb{Z}_n{}^*$. (They are distinct for exactly the same reason that they were distinct in the proof of Theorem 2.4.4.) So, these elements are all the elements of $\mathbb{Z}_n{}^*$, albeit in perhaps a different order, and in particular the products of the elements in both lists are the same. If we denote by [L] the product of the initially listed elements of $\mathbb{Z}_n{}^*$, we get an equation $[a^{\varphi(n)}]$ [L]=[L]. We can cancel the [L] – again, for exactly the same reasons as in the proof of FLT—and we arrive at $[a^{\varphi(n)}]=[1]$, which, as an equation in $\mathbb{Z}_n{}^*$, says precisely the same thing as $a^{\varphi(n)} \equiv 1$ (mod n), which is what we wanted to prove.

As an illustration of this result, suppose we want to find the remainder when 3^{43} is divided by 10. By Euler's Theorem, $3^4 \equiv 1$ (mod 10), and so $3^{40} \equiv 1$ (mod 10) as well. Then $3^{43}=3^{40}3^3$ is congruent mod 10 to $3^3=27$, which in turn is congruent mod 10 to 7. So, without any real calculation at all, we know that 3^{43} leaves a remainder of 7 when divided by 10. (Of course, even without Euler's Theorem we could tell immediately that $3^4 \equiv 1$ (mod 10) because $3^4=81$.)

Exercises

2.30 Prove the converse of Wilson's Theorem: if p is an integer greater than 1 and $(p-1)! \equiv -1$ (mod p), then p is a prime.

2.31 If p and q are distinct primes, prove that $p^{q-1}+q^{p-1} \equiv 1$ (mod pq). (Hint: Why is it enough to prove this congruence holds both mod p and mod q?)

2.32 Find the remainder when 5^{37} is divided by 14. (No electronic assistance allowed.)

2.33 Prove that 6601 is a Carmichael number. ($6601 = 7\cdot23\cdot41$)

2.34 If p is an odd prime, what is the remainder when $(p-2)!$ is divided by p?

2.6 Pythagorean Triples

Recall from high school geometry the statement of the Pythagorean Theorem: if a right triangle has side lengths x and y and hypotenuse length z, then

$$x^2 + y^2 = z^2 \tag{P}$$

Let us now shift focus a little and think of (P) not as a statement about a known entity but as an equation in the three variables x, y and z. This being a course in number theory, we are interested in solutions to this equation in positive integers. Solutions certainly exist, the simplest one being $x=3$, $y=4$, $z=5$. In this section of the book we will, using modular arithmetic as a helpful tool, study equation (P) and see if we can determine *all* solutions to this equation.

We begin with a simple observation: if (x, y, z) is a solution to (P), then so is (cx, cy, cz) for any positive integer c; after all, if $x^2+y^2=z^2$ then certainly $(cx)^2+(cy)^2=(cz)^2$. This solution being considered a somewhat trivial modification of an existing one, we focus attention on solutions that have no positive factors greater than 1 in common.

Definition 2.6.1

A *Pythagorean triple* is an ordered triple (x, y, z) of positive integers with the property that $x^2+y^2=z^2$. If the only positive integer d that divides all three of x, y and z is 1, then this is a *primitive* Pythagorean triple (hereafter denoted PPT).

If we know all PPTs, then we know all Pythagorean triples: any such triple is just obtained from a PPT by multiplying each term by a nonzero positive constant. Our ultimate goal in this section, therefore, is to classify all PPTs. We first record a simple observation about them, the proof of which is made easier by using basic properties of congruence.

Theorem 2.6.2

If (x, y, z) is a PPT, then x and y have opposite parity (i.e., one is even and the other is odd), and z is odd.

Proof. As a preliminary observation, we note that the square of an integer n is congruent to 0 mod 4 if n is even, and 1 mod 4 if n is odd. For, if n is even, it is congruent to 0 or 2 mod 4, and in either case, the result about n^2 follows. Likewise, if n is odd, it is congruent to 1 or 3 mod 4, and again the result follows.

Now, suppose first (using a proof by contradiction) that x and y are both even. In that case, it is clear that z^2 is also even, which of course implies that z is. But because (x, y, z) is a PPT, it is not possible for all three integers to be divisible by 2.

Next, suppose that x and y are both odd. In that case, by the preliminary observation above, their squares must both be congruent to 1 mod 4, which means that z^2 is congruent to 2 mod 4—which, according to the first paragraph above, can't happen.

So, x and y have opposite parity. It follows that the same is true of x^2 and y^2, which means that their sum z^2, and hence z, must be odd.

Now that we know either x or y must be odd, we will adopt the convention that x is odd, and y is even. With this convention, we can state and prove our classification theorem.

Theorem 2.6.3

If r and s are relatively prime integers of opposite parity with $r > s$, then $(r^2 - s^2, 2rs, r^2 + s^2)$ is a PPT. Moreover, any PPT (with x odd and y even) is of this form for some relatively prime integers r and s of opposite parity.

So, for example, the choice $r = 2, s = 1$ yields the familiar PPT (3, 4, 5). Another familiar example, (5, 12, 13), corresponds to $r = 3, s = 2$.

Proof of theorem 2.6.3: The easy part of the theorem is showing that any triple of the desired form is, indeed, a PPT. The fact that this triple satisfies the Pythagorean equation (P) follows immediately from some tedious, but easy, high school algebra. The fact that it is primitive follows from the easily shown fact that any prime p that divides all three terms of the PPT would have to divide either r or s; then, the fact that p divides $r^2 - s^2$ would imply that it divides both r and s, a contradiction. We leave it to the reader, as an exercise, to fill in the details.

We turn now to the less obvious half of the theorem. Starting with an arbitrary PPT (x, y, z), we must prove the existence of two relatively prime integers r and s of opposite parity such that $(x, y, z) = (r^2 - s^2, 2rs, r^2 + s^2)$.

As a first step to doing so, recall that we are assuming that x is odd and y is even. So we can write $y = 2t$ for some integer t. Substituting in equation (P) gives $4t^2 = z^2 - x^2$, or

$$t^2 = \left(\frac{z-x}{2}\right)\left(\frac{z+x}{2}\right) \tag{*}$$

Note that the two factors on the right-hand side of (*) are integers, because x and z are both odd (so their sum and difference are both even). In fact, they are relatively prime integers: if, say, a prime p divided both of them, then p

would have to divide their sum and difference, which would imply that x and z are both divisible by p. But then it would follow from (P) that p divides y as well, which contradicts the fact that (x, y, z) is a PPT.

We may now appeal to Theorem 1.5.6 to conclude that both $\left(\dfrac{z-x}{2}\right)$ and $\left(\dfrac{z+x}{2}\right)$ are squares:

$$\left(\frac{z-x}{2}\right) = s^2$$

$$\left(\frac{z+x}{2}\right) = r^2$$

Simple algebra now confirms that $(x, y, z) = (r^2 - s^2, 2rs, r^2 + s^2)$. So, to finish the proof of the theorem, we need only show that r and s are relatively prime and of opposite parity. Both of these observations, however, are practically obvious. Since $r^2 + s^2 = z^2$ is odd, it follows that r^2 and s^2, and hence r and s, must have opposite parity. And if r and s were both divisible by a prime p, then p would divide each of x, y and z, which isn't possible. So r and s are relatively prime, and the proof of Theorem 2.5.3 is complete.

There are other ways to prove this result. In Chapter 6, for example, we give a proof using the *Gaussian Integers*, an algebraic system that extends the set of (ordinary) integers.

Exercises

2.35 If (x, y, z) is a non-primitive Pythagorean triple, do there necessarily exist integers r and s (not necessarily relatively prime or of opposite parity) such that $(x, y, z) = (r^2 - s^2, 2rs, r^2 + s^2)$? Explain.

2.36 Fill in all the missing details in the proof of Theorem 2.5.3.

2.37 Find a PPT where one of x, y or z is equal to 7.

2.38 If p is an odd prime, find a PPT where $y = 4p$.

2.39 Prove that if x is any odd integer that is greater than or equal to 3, then there are integers y and z for which (x, y, z) is a PPT.

2.40 If (x, y, z) is a Pythagorean triple, prove that x or y is divisible by 3.

2.41 If (x, y, z) is a Pythagorean triple, prove that x, y or z is divisible by 5.

2.42 If (x, y, z) is a Pythagorean triple, prove that xyz is divisible by 60.

2.43 Prove that there are infinitely many PPTs (x, y, z) for which $z = y + 1$.

Challenge Problems for Chapter 2

C2.1 If a, b and n are positive integers with a relatively prime to n, and, for some integer k we have $a^k \equiv b^k$ (mod n) and $a^{k+1} \equiv b^{k+1}$ (mod n), prove that $a \equiv b$ (mod n).

C2.2 If n is a positive integer, prove that $2^{2^n} + 11$ is divisible by 3.

C2.3 If n is a positive integer, prove that $n^5/5 + n^3/3 + 7n/15$ is an integer.

C2.4 Prove that any prime $p > 5$ divides infinitely many integers, all of whose digits are 9.

C2.5 Prove that there are infinitely many positive integers n such that $n! - 1$ is composite.

C2.6 Prove that if n is composite, $\varphi(n) \le n - \sqrt{n}$, with equality when and only when n is the square of a prime.

C2.7 For positive integers m and n, prove that $\varphi(mn) = m\varphi(n)$ if and only if every prime that divides m also divides n.

C2.8 Find, with proof, all Pythagorean Triples whose entries form an arithmetic progression.

C2.9 Find, with proof, all Pythagorean Triples whose entries form a geometric progression.

3

Cryptography: An Introduction

One of the more striking "real world" applications of number theory is the study of cryptography, which concerns itself with secret communications. The ability to decipher such communications can have striking global consequences; the decipherment of the famous Zimmerman telegram (a 1917 communication from the German Foreign Office to the German ambassador in Mexico), for example, played a significant role in the decision of the United States to enter World War I. In this chapter, we give an introduction to this area of mathematics, emphasizing the role that number theory plays in it. Entire books (e.g., [R-S]) have been written on this subject, so our chapter-long treatment of it will necessarily hit only a few high points. Our approach will be somewhat informal; we will focus on the mechanics of the applications rather than excessive mathematical formalism.

3.1 Basic Definitions

We begin with some terminology. Our basic set up is as follows: one person (traditionally referred to as "Alice") wishes to send a message to another person ("Bob"), but both wish to keep the contents secret from adversarial eavesdropper third parties (collectively referred to as "Eve"). Alice will attempt to do this by converting the original (*plaintext*) message to a secret or disguised one, the *ciphertext*. The process of converting the plaintext message to a ciphertext one is called *encryption* or *enciphering*; the reverse process, employed by Bob to translate the secret message back to its original form, is called *decryption* or *deciphering*. The study of methods for converting plaintext to ciphertext is called *cryptography*; the study of deciphering secret messages is called *cryptanalysis*. The term *cryptology* is used to embody the study of both cryptography and cryptanalysis; practitioners of this study are *cryptologists*. A *cryptosystem* or *cipher* is a particular method of cryptography.

To illustrate these ideas, let us start with a simple-minded cryptosystem that does not involve much mathematics at all. Suppose we begin by writing out the letters of the alphabet in order:

$$A\ B\ C\ D...Z$$

DOI: 10.1201/9781003318712-4

And underneath that, write out some permutation (rearrangement) of the letters. For example, we can write the letters in reverse order:

$$A\ B\ C\ D...Z$$

$$Z\ Y\ X\ W...A$$

To encrypt a message, simply take every letter of the plaintext and replace it by the letter that appears underneath it in the list above. For example, BAD converts to YZW. To decrypt the message, we of course just replace every letter in the ciphertext by the letter that appears immediately above it.

The disadvantages of this "substitution cipher" should be obvious. First, if we try to maximize security by taking a purely random rearrangement of the alphabet, we run the risk of having to either memorize the substitution key (a daunting chore) or writing it down somewhere, which leads to the possibility of the writing being stolen or otherwise accessed. Also, this kind of substitution cipher can be broken by *frequency analysis*. There are tables that list the relative frequency of all letters (and two-letter couplets, etc.) in the English alphabet. The top ten most frequently occurring letters in the English language are, in decreasing order of frequency: E, T, A, O, N, R, I, S, H and D. So, in a message of any significant size, one might look for the letters that appear the most often and make guesses as to what they are. There are also tables that list the relative frequency of two-letter and three-letter groupings, and frequency analysis can be applied here as well. Once a number of letters have been filled in, the rest of the message can often be deciphered by figuring out what makes sense.

Since the substitution method doesn't seem very useful in practice, let us bring mathematics into the picture and see how number theoretic ideas can be used.

3.2 Classical Cryptography

We begin with a cryptosystem that dates back to Julius Caesar and is referred to as the *Caesar cipher*: we simply advance every letter in the plaintext three letters forward (so that A, for example, would become D); for the letters X, Y and Z, we just "wrap around" to the beginning of the alphabet, so that X becomes A, Y becomes B and Z becomes C. So, for example, if Alice wanted to send Bob the message BRUTUS, she would send EUXWXV.

The idea of "wrapping around" to the beginning of the alphabet suggests modular arithmetic, and indeed it is easy to see how to formalize this mathematically. If we identify every letter with a number representing its position in the alphabet (starting, for convenience, with 0 rather than 1), we see that we can identify the letter A with 0, B with 1 and so on, until we get to Z, which is assigned number 25. Encryption under the Caesar cipher simply amounts to then applying the function

$$x \rightarrow x + 3 \ (\text{mod } 26)$$

and decryption amounts to applying the function

$$x \rightarrow x - 3 \ (\text{mod } 26)$$

Of course, there is nothing magical about the number "3"; it can be replaced by any other integer between 1 and 25. This is an example of a *shift cipher*— we simply shift each letter a certain number of places.

As a cryptosystem, though, shift ciphers have serious defects. For one thing, if Eve knows the method Alice and Bob are using (and it is good practice to assume your adversary does know the method, if not the key; this is called *Kerchkoff's principle*), then all Eve has to do is check 25 numbers and see which one works. This can easily be done by hand, let alone in seconds with a computer.

Another problem here is that this method is also subject to attacks by frequency analysis. This is not surprising, since this method is really just an example of a substitution cipher, with the substitutions being given by a specific formula.

There is a mild generalization of shift ciphers that, unfortunately, are still vulnerable to the kinds of attacks mentioned above. This is the idea of an *affine cipher*. Whereas a shift cipher deals with functions of the form

$$x \rightarrow x + b (\text{mod } 26),$$

affine ciphers deal with functions of the form

$$x \rightarrow a x + b (\text{mod } 26)$$

Of course, to ensure that different plaintext letters get mapped into different ciphertext letters (this is necessary if we want to ensure that Bob can invert the process), we must ensure that this function is 1-1 and onto. For this to happen, a must be invertible modulo 26. This means that the integer a must be relatively prime to 26; the total number of distinct (mod 26) integers that satisfy this condition is $\varphi(26) = 12$.

As an illustration, consider the affine cipher given by the key

$$x \rightarrow 5x + 9 \ (\text{mod } 26)$$

Suppose Bob receives the encrypted message JIIXWD. The letter J (the 10th letter of the alphabet) corresponds to the number 9, so to find the plaintext letter that corresponds to it, Bob must solve (modulo 26) $9=5x+9$, or $0=5x$. The only solution to this is $x=0$, so A is the letter that encrypts to J. Proceeding in a similar fashion through the remaining letters of the word, Bob eventually arrives at the word AFFINE.

However, just as is the case with shift ciphers, affine ciphers are also vulnerable to quick searches through all possible values of a and b, as well as to frequency analysis attack. If Eve can identify the images of two letters, she can set up a system of linear equations that can be solved modulo 26. Suppose, for example, that having intercepted a message from Alice to Bob and knowing that an affine cipher is being used, Eve also knows, or at least guesses, that E encrypts to B and that T encrypts to A. Eve can then write down two equations:

$$1 \equiv 4a + b$$

$$0 \equiv 19a + b$$

(All congruences are mod 26, of course.) Subtracting the first from the second, we get

$$1 \equiv -15a$$

or

$$1 \equiv 11a$$

from which we conclude that $a=19$, since $11 \times 19 = 209 \equiv 1 \pmod{26}$. Now that we know $a=19$, substituting in the first equation gives $b \equiv -75 \equiv 3 \pmod{26}$. So our affine cipher is

$$x \rightarrow 19x + 3 \pmod{26},$$

and now (if Eve was originally guessing) all she has to do is check to see if this works.

One way to help foil a frequency attack is to use a cryptosystem in which a letter in the plaintext does not necessarily always correspond to the same letter in the ciphertext and vice versa. For example, we might come up with a system in which the letter A in the plaintext is encrypted to C on one occasion and to X on another. There are several such systems known. Some are called *block ciphers* because they operate on blocks of letters rather than individual letters. The first of these that we will discuss dates back to the 16th century and is known as the *Vigenere cipher*.

This works as follows. Alice and Bob agree on a word (say, for purposes of illustration, DOOR). This word happens to have four letters in it, so the cipher will operate on blocks of four letters each (or fewer, if we run out of letters). Translating the word DOOR into its numerical equivalent, we get a four-component vector (3, 14, 14, 17). Now, given any plaintext message (say, RETREAT), our first step is to break it into four-word blocks; since RETREAT has seven letters, our second block will only have three letters. We get RETR and EAT. Now, in the first block, we shift R by 3, E by 14, T by 14 and R by 17. This gives us USHI. In the second block, we shift E by 3, A by 14 and T by 14, giving us HOH. Alice will encrypt her message, therefore, as USHIHOH. The most frequently occurring letter in the ciphertext is H, but this tells us nothing, because the first H corresponds to T, the second to E and the third to T.

When Bob gets the ciphertext, he can easily use modular arithmetic to decrypt it. He must simply subtract instead of add, or, to put it another way, he must add 23, 12, 12, and 9 (since these numbers are, respectively, the additive inverses (modulo 26) of 3, 14, 14 and 17).

We should point out that, in practice today, nobody really uses the Vigenere cipher. Although a straightforward frequency analysis does not work, the cipher is vulnerable to other kinds of statistical attack, but discussing this would take us too far afield. Suffice it to say, though, that longer key words (or phrases) are more secure than shorter ones.

Another example of a block cipher uses arithmetic modulo 26 applied to matrices and is called the *Hill cipher*. (This discussion assumes familiarity with matrix multiplication.) Here, Alice and Bob agree beforehand on a square matrix whose entries are integers modulo 26. (We could work with integers modulo n, but 26 is customary; it is also common to work with 2×2 matrices, and we shall do so in what follows.) In order to be invertible, the determinant of the matrix must be a unit modulo 26—i.e., must be relatively prime to 26. To illustrate the procedure, we will take

$$A = \begin{pmatrix} 7 & 2 \\ 5 & 3 \end{pmatrix}$$

(which is invertible, since the determinant of A is 11, which is relatively prime to 26). Now let us take, as our plaintext message, the word MONEY. We break this up into chunks of size 2 (the order of the matrix) or fewer: MO NE Y. The first chunk corresponds to the vector $x = (12, 14)$. To encrypt it, we simply use matrix multiplication and compute xA, which a short calculation reveals to be (24, 14). This corresponds to the letters Y and O. Next, we deal with NE and compute the product of the vector (13, 4) with the matrix A, getting (7, 12) or HM. Finally, to deal with the remaining Y, we extend this to a pair by adding the letter Z, resulting in the vector (24, 25). The product of this with A gives (7, 19), or HT. So our ciphertext is YOHMHT.

Decryption follows the same pattern, but we use the matrix inverse $A^{-1} = \begin{pmatrix} 5 & 14 \\ 9 & 3 \end{pmatrix}$. If $y = xA$, then $x = yA^{-1}$. Bob will take the ciphertext YOHMHT and break it into chunks of two letters each: YO HM HT. The first group of letters corresponds to $(24, 14)A^{-1}$, which (of course) is $(12, 14)$. Proceeding in this way, Bob deciphers the ciphertext and arrives at MONEYZ. He ignores the placeholder Z and gets MONEY as his secret message.

We should remark that some authors write vectors **x** as column vectors and therefore compute A**x** rather than **x**A. This is a matter of personal preference; all that matters is that Alice and Bob have an understanding as to what approach is being used.

Exercises

3.1 Use the Caesar cipher to encrypt the plaintext LAUNCH THE ATTACK.

3.2 Use the Caesar cipher to decrypt the ciphertext QHYHU YDB DQBWKLQJ.

3.3 Use the affine cipher $x \to 5x+9 \pmod{26}$ to decrypt ABG YDTQDA.

3.4 For the affine cipher above, encrypt FOLLOW THE MONEY and DROP ALL BOMBS.

3.5 Use the affine cipher above to decrypt PS RL HZ IW HZ.

3.6 Use the Vigenere cipher given in the text to encrypt I LOVE YOU.

3.7 Use the Hill cipher given in the text to encrypt SURRENDER.

3.8 (This interesting example comes from [KW].) It was noted in the text that for an affine cipher $x \to ax+b \pmod{26}$ to work, a must be relatively prime to 26. This example illustrates why. Consider the affine cipher $x \to 13x+5 \pmod{26}$ and the ciphertext message SFSF. Show that this message can be decrypted in two ways, yielding plaintext messages that are diametrically opposite in meaning.

3.3 Public Key Cryptography: RSA

The cryptosystems that were studied in the previous section all depended on a *key*—i.e., some information known to Alice and Bob, but nobody else. For affine ciphers, the key consisted of the two integers a and b; for the Vigenere cipher, it consisted of the key word; for the Hill method, it consisted of a matrix. If Eve, the enemy, were to become aware of this key (by, say, spying), then all would be lost: she could read the messages as easily as Alice and Bob could.

For this reason, attention turned in the 20th century to cryptosystems that did not depend on secret information like this. In this section, we will discuss one of these methods, known as RSA (named for the discoverers Rivest, Shamir and Adleman). As in the previous section, we are more interested in an explanation of how the method works and what it has to do with modular arithmetic than with a rigorous development of the subject, so we will dispense with mathematical formalisms.

Essentially, the RSA method depends for its validity on this rule of thumb: *multiplying is easy, factoring is hard.* By this, we mean that it is computationally easy to multiply two integers (even two very large ones), but, given a very large number, there is currently no known computationally fast algorithm for factoring it. (If somebody were to discover one, the whole face of number theory, as well as government and modern business, would be dramatically changed.)

The basic idea is as follows. Alice has a message that she wants to send to Bob; since words can be converted to integers, we assume the message is a number, say m. Bob begins by selecting two distinct prime integers p and q (in practice, both very large) and then multiplying them together to form the integer $n = pq$. He also computes $\varphi(n) = (p-1)(q-1)$. These computations can be done easily on a computer, and once they are done Bob has no further need of the individual primes p and q; he does not even need to let Alice know what they are. Bob does transmit to Alice what the number n is, but he does not need to worry that this message might be intercepted: the beauty of the method is that the whole world can know what n is as well; there is no need to keep it secret.

Bob also selects an integer e that is relatively prime to $\varphi(n)$ and transmits this number to Alice as well. Alice then computes m^e (mod n); this is the ciphertext (call it c) that is transmitted by Alice to Bob. (This computation can be done quickly on a computer; in fact, there are modular arithmetic calculators available online that instantly produce an answer when the base, exponent and modulus are entered.)

By the way e was selected, it has a multiplicative inverse modulo $\varphi(n)$; call it d. To decrypt the message, Bob simply computes c^d (mod n), which can also be done quickly on a computer. Euler's Theorem guarantees that Bob's computation recovers the number m. To see why, observe that (by definition of multiplicative inverse mod $\varphi(n)$), $ed = k\,\varphi(n)+1$ for some integer k. Thus, with all calculations below being done modulo n, we have

$$c^d \equiv \left(m^e\right)^d$$

$$= m^{ed}$$

$$= m^{k\varphi(n)+1}$$

$$= m^{k\varphi(n)}m$$

$$= \left(m^{\varphi(n)}\right)^{k} m$$

$$\equiv 1^{k} m \left(\text{by Euler's theorem}\right)$$

$= m$. So, Bob has recovered the original plaintext message.

We illustrate this with an example. Since we want to illustrate the method and not get lost in lengthy calculations, we will use simple numbers that are ridiculously smaller than would ever be used in real life. (Small numbers are deadly to effective use of the RSA method because a small n can be easily factored, and once Eve knows what p and q are, the game is essentially lost.) Suppose, then, that Bob selects $p=29$, $q=43$, so that $n=1427$ and $\varphi(n)=1176$. The number 5 is relatively prime to 1176, so take $e=5$. A calculation then shows that the multiplicative inverse d of e is 941.

Suppose the message Alice wants to send to Bob is SEND MONEY. Translating each letter into its numerical equivalent, this results in the string of integers

18   04   13   03   12   14   13   04   24. Using the publicly known numbers n and e, Alice then computes, modulo n, the 5th power of each of these listed numbers. She obtains

363   1024   934   243   679   367   934   1024   529. This is the message that she transmits to Bob, who, in turn, then computes the 941st power of each of these numbers modulo n, recovering the initial string of numbers, which immediately translates back to the message SEND MONEY.

Exercises

In doing these exercises, you can use the Repeated Squaring Method discussed earlier, or a computer, or an online modular arithmetic calculator such as the one found at http://ptrow.com/perl/calculator.pl

3.9   Encrypt I LOVE YOU using $p=7, q=13, e=5$. How will Bob decrypt this?

3.10   Encrypt SETTLE THE CASE using $p=5, q=17, e=3$. How will Bob decrypt this?

3.11   Would it make any sense at all to select $e=1$? Explain.

3.12   Will 2 ever be selected as the encryption exponent e? Explain.

Challenge Problems for Chapter 3

C3.1 Give an example of two English words, each with at least four letters, that encrypt to each other under a shift cipher.

C3.2 You are Eve and have intercepted the message RSGVZ F TRMM-RFESRO HYEDSX IERFJ from Alice to Bob. You know they are using an affine cipher but don't know the key. Decrypt the message without using technology.

C3.3 Suppose that you are given $n = pq$, a product of two very large distinct primes p and q. Although it is not computationally easy to determine p and q, it *is* easy to do so if you are also given φ (n). Explain why. (*Hint*: p and q are the roots of the polynomial $(x - p)(x - q)$; explain how the given information allows you to find the coefficients of this polynomial.)

4

Perfect Numbers

The subject matter of this chapter, perfect numbers, has ancient roots, dating back to the time of Euclid, where they are discussed in Book IX of his monumental treatise *The Elements*. Interestingly, however, only four specific perfect numbers were actually known to the Greeks. Now we know many more, but as we will see there are still several long-standing open questions concerning them.

4.1 Basic Definitions and Principles: The Sigma Function

Perfect numbers are easy to define:

Definition 4.1.1

A positive integer $n>1$ is called *perfect* if the sum of the positive divisors of n, other than n itself, is equal to n. Equivalently, $n>1$ is perfect if the sum of *all* of the positive divisors of n is equal to $2n$.

The two smallest perfect numbers are 6 and 28, as one can easily verify:

$$1 + 2 + 3 = 6$$

$$1 + 2 + 4 + 7 + 14 = 28.$$

Because we will be dealing with the sum of the positive divisors of a positive integer, it is convenient to adopt a compact notation. Accordingly, we have:

Definition 4.1.2

If n is a positive integer, then the sum of all divisors of n (including n itself) is denoted $\sigma(n)$. The function σ that is defined by this is known as the "σ function".

So, perfect numbers are precisely those integers $n>1$ for which $\sigma(n)=2n$. Also, $\sigma(p)=p+1$ if and only if p is a prime. As another example, $\sigma(10)=18$ (the sum of 1, 2, 5 and 10). And of course $\sigma(1)=1$.

DOI: 10.1201/9781003318712-5

This is not the first time that we have defined a function from the set of positive integers into the set of positive integers; the Euler phi function of Section 2.4 is another example of one. It turns out that both the σ and φ functions have an important property, which we define next.

Definition 4.1.3

A function f from the set of positive integers into the set of positive integers is called *multiplicative* if $f(mn)=f(m)f(n)$ whenever m and n are relatively prime positive integers.

We have previously proved (Theorem 2.4.4) that the φ function is multiplicative; we now show that the σ function is as well.

Theorem 4.1.4

The σ function is multiplicative.

Proof. Suppose that m and n are relatively prime positive integers. We must show that $\sigma(mn)=\sigma(m)\sigma(n)$. We may assume without loss of generality that both m and n are greater than 1, as the result is obvious otherwise. To do this, we first think about how the divisors of mn relate to the divisors of m and the divisors of n. (Throughout this proof, we will use "divisor" to mean "positive divisor".) Think of m as being written as a product of prime powers in a unique way, and let $\{p_i\}$ be the set of primes appearing in this factorization. Similarly, n can be written as a product of prime powers in a unique way as well, so let us denote by $\{q_j\}$ the set of primes appearing in this factorization. Since m and n are relatively prime, there is no overlap among these two sets of primes. Now, by the uniqueness of factorization, any divisor of mn must consist of a product of the p_i and q_j (perhaps with zero exponent); by lumping together the p's and the q's, we see that any divisor of mn can be written uniquely as de, where d is a divisor of m and e is a divisor of n. Conversely, any number of the form de, where d is a divisor of m and e is a divisor of n, is obviously a divisor of mn.

So, if we denote the divisors of m by $d_1, \ldots, d_r$ and the divisors of n by $e_1, \ldots, e_s$, then $\sigma(mn)$ is the sum of all $d_i e_j$, $\sigma(m)$ is the sum of all d_i, and $\sigma(n)$ is the sum of all e_j. This makes it clear that $\sigma(mn)=\sigma(m)\sigma(n)$, and the proof is complete.

We can also describe how the function σ treats prime powers. The simple observation here is that if p is a prime and k is a positive integer, then the only divisors of p^k are the prime powers p^j where $0 \le j \le k$. Thus, by definition, $\sigma(p^k) = 1+p+\ldots+p^k$, which, by the formula for the sum of a finite geometric progression, is $(p^{k+1} - 1)/(p - 1)$. We have thus proved the following theorem.

Theorem 4.1.5

If p is a prime and k is a positive integer, then $\sigma(p^k)=1+p+\ldots+p^k=(p^{k+1}-1)/(p-1)$.

Combining the two previous theorems gives us a nice way to compute $\sigma(n)$, *if* we know the prime factorization of n. For example, $\sigma(100)=\sigma(2^2)\ \sigma(5^2)=7\cdot31=217$. In the next section, we will use properties of the sigma function to obtain a characterization of all even perfect numbers. Although this characterization is interesting, it does not answer the following question about even perfect numbers, which remains an unsolved problem to this day:

Are there infinitely many even perfect numbers?

For odd perfect numbers, even less is known. Indeed, nobody even knows the answer to this question:

Are there any odd perfect numbers at all?

Exercises

4.1 Compute, without electronic assistance, $\sigma(1821)$ and $\sigma(12060)$.

4.2 For which positive integers n is it true that $\sigma(n)=n$? Explain.

4.3 For which positive integers n is it true that $\sigma(n)=n+1$? Explain.

4.4 Prove that a perfect number can never be prime.

4.5 Prove that a perfect number can never be a prime power.

4.6 Let f be a multiplicative function. Define a new function F on the set of positive integers as follows: $F(n)=\sum f(d)$, where the sum ranges over all positive divisors d of n. Prove that F is a multiplicative function.

4.7 Explain why Theorem 4.1.4 is a special case of the previous exercise.

_____

4.2 Even Perfect Numbers

Our goal in this section is to obtain a characterization of even perfect numbers. We begin by looking at a special kind of prime number (a Mersenne prime, named after Marin Mersenne, a 17th-century French friar) that, on its face, seems to have nothing to do with perfect numbers, but which ultimately play an important role in their classification.

Definition 4.2.1

A *Mersenne prime* is a prime of the form 2^n-1 for some integer n.

So, for example, $3=2^2-1$, $7=2^3-1$ and $31=2^5-1$ are all Mersenne primes. Notice that in these cases the exponent n is itself a prime. This is not an accident, as the next theorem shows.

Theorem 4.2.2

If, for some positive integer n, $2^n - 1$ is a prime, then n is a prime.

Proof. Clearly n must be greater than 1. If n is not a prime, then it has a nontrivial factorization $n = ab$. But this then results in a nontrivial factorization of $2n - 1$:
$2ab - 1 = (2a - 1)(1 + 2a + 22a + \ldots + 2(b - 1)a)$. This contradicts the fact that $2n - 1$ is a prime.

We caution the reader that the previous theorem is not an "if and only if" one. If n is prime, $2^n - 1$ need not be, as Exercise 5.9 shows. Nobody knows, in fact, whether there are infinitely many Mersenne primes.

Our next theorem establishes a connection between Mersenne primes and perfect numbers.

Theorem 4.2.3

If $2^p - 1$ is a Mersenne prime, then $n = 2^{p-1}(2^p - 1)$ is a perfect number.

Proof. Note that 2^{p-1} and $2^p - 1$ are relatively prime. Therefore, $\sigma(n) = \sigma(2^{p-1}(2^p - 1)) = \sigma(2^{p-1})\ \sigma(2^p - 1)$. Now, $\sigma(2^{p-1}) = 2^p - 1$ by Theorem 4.1.5, and $\sigma(2^p - 1) = 2^p$ because $2^p - 1$ is a prime. Thus, $\sigma(n) = \sigma(2^{p-1})\ \sigma(2^p - 1) = 2^p(2^p - 1) = 2n$, and n is perfect.

This theorem has been known for thousands of years; it appears in Euclid's *Elements*. What is less easy to show is that the converse of this result is true; this result was proved by Euler in the 18th century. That is the main characterization of even perfect numbers that we are looking for. There are a number of different proofs of this result; we give Euler's proof.

Theorem 4.2.4

If n is an even perfect number, then $n = 2^{p-1}(2^p - 1)$ for some Mersenne prime $2^p - 1$.

Proof. Because n is even, we can write $n = 2^k m$ for some positive integers k and m, with m odd. Note that m must be greater than 1, since a power of 2 is never a perfect number by Exercise 4.5. We will show that m is a Mersenne prime $2^p - 1$ and that $k = p - 1$, thus proving the result.

Because n is perfect and 2^k is relatively prime to m, we have

$$2^{k+1}m = 2n = \sigma(n) = \sigma(2^k)\ \sigma(m) = (2^{k+1} - 1)\sigma(m). (*)$$

From this, it follows that $2^{k+1} - 1$ divides $2^{k+1}m$, which in turn implies that $2^{k+1} - 1$ divides m by Theorem 1.3.7. Let us see what this entails. If we write

$$m = (2^{k+1} - 1)\ r, (**)$$

substitute above, and then divide by $2^{k+1} - 1$, we get

$$2^{k+1}r = \sigma(m). \text{ (***)}$$

Now r is a divisor of m based on the way it was defined, and of course m is a divisor of m. So $\sigma(m)$, the sum of all divisors of m must be at least as big as $m+r$.

Once we have this, we can show that we actually have equality, thanks to this chain of equalities and inequalities: $\sigma(m)$

$$\geq m+r$$

$$= 2^{k+1} \; r \text{ (from **)}$$

$$= \sigma(m). \text{ (from ***)}$$

Since $\sigma(m)$ is the sum of *all* divisors of m and is also the sum of the two divisors m and r, it follows that these are all the divisors of m. But one divisor of m is, of course, 1. So we must have $r=1$ and $\sigma(m)=m+1$. This means (Exercise 4.3) that m is a prime. By Theorem 4.2.2 and (**), this means that $k+1$ is a prime. If we write $k+1=p$, then

$n=2^k m = 2^p - 1 \, (2^p - 1)$, and the proof is complete.

The case $p=2$ in this theorem yields $n=6$, and the case $p=3$ gives the even perfect number 28. As an exercise, check for other even perfect numbers using other values of p.

Exercises

4.8   Theorem 4.2.2 says that $2^{10} - 1$ can't be a prime. Verify this directly.

4.9   Verify without electronic assistance that $2^{11} - 1$ is not prime, thus showing that the converse of Theorem 4.2.2 is false.

4.10 Find two more even perfect numbers.

4.11 A *triangular number* is one that can be written as the sum of the first n positive integers, for some n. Prove that any even perfect number is triangular.

4.12 Prove that the unit digit of any even perfect number is either 6 or 8.

Challenge Problems for Chapter 4

C4.1 If m and n are prime powers (not necessarily of the same prime) and $\sigma(n)/n = \sigma(m)/m$, prove that $m=n$.

C4.2   If m and n are positive integers, and m divides n, prove $\sigma(n)/n \geq \sigma$ $(m)/m$, with equality when and only when $m=n$. Use this to prove that a proper divisor of a perfect number is never perfect.

C4.3   If n is a perfect number, what is the sum of the reciprocals of all positive divisors of n? Use this to give an alternative proof that a proper divisor of a perfect number is never perfect.

C4.4   Let $m>2$. Prove that n satisfies $(m-1)>$ $(n)=mn$ if and only if $n=m-1$ and is a prime.

5

Primitive Roots

The subject matter of this chapter has strong algebraic content and is, in fact, largely a special case of basic group theory in abstract algebra. Of course, abstract algebra is not a prerequisite for this text, so the material will be developed from scratch in a number-theoretic setting. However, since it would be a pity not to understand this material in its proper context, occasional references to group theory will be made. These references can be ignored by people who are not interested in these algebraic connections.

5.1 Order of an Integer

If n is a positive integer and a is an integer that is relatively prime to n, then Euler's theorem (Theorem 2.4.6) tells us that

$$a^{\varphi(n)} \equiv 1 (\bmod n)$$

In particular, some positive power of a is congruent to 1 mod n. It follows by Well-Ordering that there is a *smallest* positive power of a with this property. That observation motivates the following definition.

Definition 5.1.1

If n is a positive integer and a is an integer that is relatively prime to n, then the *order of a mod n*, denoted $\mathrm{ord}_n(a)$, is the smallest positive integer k with the property that $a^k \equiv 1 (\bmod n)$.

So, for example, the order of 3 mod 10 is 4, as we can see by repeated exponentiation: $3^4 = 81$, which is congruent to 1 mod 10, and this is the first positive power of 3 with that property. On the other hand, the order of 9 mod 10 is 2. The order of 4 mod 10 is not defined, because 4 and 10 are not relatively prime.

It is obvious (why?) that if a and b are congruent mod n, and both relatively prime to n, then $\mathrm{ord}_n(a) = \mathrm{ord}_n(b)$. Recall that $a^k \equiv 1 (\bmod n)$ if and only if $[a]^k = [1]$ in $\mathbb{Z}_n^*$. Thus, we can characterize the order of the integer a as the smallest positive integer k with the property that $[a]^k = [1]$ in $\mathbb{Z}_n^*$, which, in

DOI: 10.1201/9781003318712-6

Section 2.2, we defined to be the set of units of $\mathbb{Z}_n$. Here is where algebra enters the picture: in Section 2.2, we showed that this set was an algebraic structure called a *group*. In this context, the positive integer k is called the *order* of the element $[a]$ in the group $\mathbb{Z}_n^*$.

Theorem 5.1.2

If the order of a mod n is k, then no two of the integers $a, a^2, ..., a^k$ are congruent mod n. Equivalently, in the group $\mathbb{Z}_n^*$, the elements $[a], [a]^2, ..., [a]^k$ are all distinct.

Proof. Suppose to the contrary that a^i and a^j are congruent mod n, where $0 < i < j \le k$. This yields $a^{i-j} \cdot a^j \equiv a^j \pmod{n}$. Since a^j is relatively prime to n (why?), we can cancel it in this congruence equality, getting $a^{i-j} \equiv 1 \pmod{n}$, where $i-j$ is a positive integer less than k. This contradicts the fact that the order of a mod n is k.

If k is the order of a mod n, and d is any integer such that $a^d \equiv 1 \pmod{n}$, then of course k must be less than or equal to d. But our next theorem says even more—namely, that k must divide d. The proof uses a familiar technique.

Theorem 5.1.3

If k is the order of a mod n, and d is any integer such that $a^d \equiv 1 \pmod{n}$, then k divides d.

Proof. The Division Algorithm tells us that we can write $d = kq + r$, where $0 \le r < k$. From here it is obvious that $1 \equiv a^d \equiv a^r \pmod{n}$. This can only happen if $r = 0$, as otherwise r would be a positive integer, less than k, which gives 1 when a is raised to that power, contradicting the fact that k is the order of a. So $r = 0$, and $d = kq$. In particular, k divides d.

Combining the previous result with Euler's Theorem immediately gives us:

Corollary 5.1.4

If k is the order of a mod n, then k divides $\varphi(n)$.

This corollary should not, of course, come as a surprise to anybody familiar with elementary group theory: the order of an element of a group always divides the order of the group, by Lagrange's Theorem. (See Appendix C.) Applying this to the group $\mathbb{Z}_n^*$, which has order $\varphi(n)$, immediately gives the result.

We end this section with one other result about the order of an integer, which can be roughly summed by saying "an integer and its multiplicative inverse mod n have the same order mod n". The fairly simple proof of this result is left as an exercise. People who have taken abstract algebra will note that this result, too, is really a special case of basic group theory.

Theorem 5.1.5

If n is a positive integer and a is an integer that is relatively prime to n, and b is an integer with the property that $ab=1 \pmod{n}$, then $\text{ord}_n(a)=\text{ord}_n(b)$.

Exercises

5.1 Prove Theorem 5.1.5.

5.2 What integers between 1 and n have order 1 mod n?

5.3 Find the order of 5 mod 12.

5.4 Find the order of 3 mod 7.

5.5 Find an integer that has order 10 mod 11.

5.2 Primitive Roots

If the integer a is relatively prime to the positive integer n, then we know that the order of a modulo n cannot be greater than $\varphi(n)$. If a has this maximal order, it is called a *primitive root* of n:

Definition 5.2.1

If n is a positive integer, a *primitive root* of n is an integer a, relatively prime to n, whose order mod n is $\varphi(n)$.

Primitive roots need not exist. For example, take $n=8$. $\mathbb{Z}_8^* = \{[1],[3],[5],[7]\}$, and it is easy to see that every element of this set squares to [1]. Thus, every integer has order 1 or 2 mod 8; there is no integer of order $4=\varphi(8)$. We will see later, however, that if p is a prime, a primitive root mod p always exists. This is the ultimate objective of this chapter.

It follows from Theorem 5.1.2 that if a is a primitive root mod n, then every element of the set $\mathbb{Z}_n^*$ is a power of $[a]$. This is because the powers $[a], ..., [a]^{\varphi(n)}$ are all distinct, and because there are $\varphi(n)$ such powers, they must constitute all the $\varphi(n)$ elements of $\mathbb{Z}_n^*$. Conversely, if the powers $[a], ..., [a]^{\varphi(n)}$ are all distinct, then a must have order $\varphi(n)$ (why?). This is the true significance of an integer being a primitive root, and it, too, is a statement about algebra: the existence of a primitive root mod n says that $\mathbb{Z}_n^*$ is a *cyclic* group. We summarize this discussion in a theorem:

Theorem 5.2.2

The integer a is a primitive root modulo n if and only if every integer that is relatively prime to n is congruent, mod n, to some power of a, which in turn happens if and only if every element of $\mathbb{Z}_n^*$ is a power of $[a]$.

Let's consider the example $n=5$. We have $\mathbb{Z}_5^* = \{[1],[2],[3],[4]\}$, and a simple calculation shows that $[2]^1=[2]$, $[2]^2=[4]$, $[2]^3=[3]$, $[2]^4=[1]$. Thus, the smallest power of 2 that is congruent to 1 mod 5 is the 4th power. Since $4=\varphi(5)$, 2 is a primitive root mod 5. At the same time, we see that the powers of [2] sweep out all the elements of $\mathbb{Z}_5^*$, as we would expect from the previous theorem.

Primitive roots, when they exist, are generally not unique. In the previous example, we know that 3, the multiplicative inverse of 2 mod 5, must also have order 4 (see Theorem 5.1.5) and so must also be a primitive root mod 5. We can verify this directly by computing powers of [3] and observing that we don't get to [1] until the fourth power: $[3]^1=[3]$, $[3]^2=[4]$, $[3]^3=[2]$, $[3]^4=[1]$. The following theorem describes explicitly how one primitive root is related to another.

Theorem 5.2.3

If a is a primitive root modulo n, and k is a positive integer, then a^k is also a primitive root modulo n if and only if k and $\varphi(n)$ are relatively prime.

Proof. Suppose first that k and $\varphi(n)$ are relatively prime. Let d denote the order of a^k mod n. Then $a^{kd} \equiv 1 \pmod{n}$, from which we conclude (Theorem 5.1.3) that $\varphi(n)$ divides kd. But since k and $\varphi(n)$ are relatively prime, this implies that $\varphi(n)$ divides d. But we also know that d divides $\varphi(n)$, so this gives $d=\varphi(n)$. This proves that a^k is also a primitive root, as desired.

For the converse, assume that a^k is a primitive root. Suppose, hoping for a contradiction, that k and $\varphi(n)$ are not relatively prime; let us denote by d their greatest common divisor, and note that $d>1$. Next, note that $a^{k\varphi(n)/d} \equiv 1 \pmod{n}$ (why?). This tells us that a^k has order at most $\varphi(n)/d<\varphi(n)$, contradicting the fact that a^k is a primitive root. This concludes the proof.

To illustrate the ideas of this section, let us consider the following problem: *Find all primitive roots (if any) mod 11.* This question is identical to the question of finding all elements in $\mathbb{Z}_{11}$ that have order 10. We could do this by taking every element [1], ..., [10] and computing powers, but let's see if we can do this more systematically, without resorting to such tedious computation. First of all, let us consider the order of 2 mod 11. We know from our work in this chapter that it must be a divisor of 10. In other words, it must be either 1, 2, 5 or 10. But it is easy to see that in $\mathbb{Z}_{11}$ we have $[2]^1=[2]$, $[2]^2=[4]$ and $[2]^5=[10]$. Since none of these powers are equal to [1], the order of [2] cannot be 1, 2 or 5. Hence, it must be 10. In other words, 2 is a primitive root mod 11. By the previous theorem, all the primitive roots are of the form 2^k where k is relatively prime to 10. So, working in $\mathbb{Z}_{11}$, the elements of order 10 are [2], [8] (=$[2]^3$), [7] (= $[2]^7$) and [6] (= $[2]^9$).

An interesting question, motivated by the example above, is: are there infinitely many primes that have 2 as a primitive root? The conjecture that there are, known as *Artin's Conjecture*, is yet another example of an unsolved problem in number theory.

Our next goal is to prove that primitive roots exist for any prime integer p. To accomplish this task, we need to study polynomials, which we do in the next section.

Exercises

5.6   Determine, with justification, whether 2 is a primitive root mod 13.

5.7   Given that there exists a primitive root mod 101, how many (mod 101) are there?

5.8   Given that there exists a primitive root mod n, how many (mod n) are there?

5.9   Find, with proof, all positive integers n that have exactly one primitive root mod n.

5.10 If p is a prime of the form $4k+1$, prove that g is a primitive root mod p if and only if $-g$ is.

5.3 Polynomials in $\mathbb{Z}_p$

Throughout this section, p denotes an arbitrary but fixed prime. We will be working a lot with the set $\mathbb{Z}_p = \{[0],[1],\ldots,[p-1]\}$. Because of this, for notational convenience, we will abuse notation and drop the brackets on the elements of this set, writing them to look like integers. But we must keep in mind that they are not integers, and that addition and multiplication are all done modulo p. For example, in $\mathbb{Z}_7$, $3+6=2$.

In high school, you undoubtedly learned about polynomials with real coefficients, so let us very briefly recall some of the relevant facts about them. A *polynomial with real coefficients* is an expression $f(x) = a_0 + a_1x + \cdots + a_nx^n$, where the a_i are all real numbers; they are called the *coefficients* of the polynomial $f(x)$. If the a_i are not all zero, then the largest n for which a_n is nonzero is called the *degree* of $f(x)$.

Polynomials can be added and multiplied. Addition is particularly simple: we simply add "like terms". In other words, the if $f(x) = a_0 + a_1x + \cdots + a_nx^n$ and $g(x) = b_0 + b_1x + \cdots + b_mx^m$, then the coefficient of x^k in $f(x)+g(x)$ is simply a_k+b_k. Multiplication is slightly more complicated: with $f(x)$ and $g(x)$ as before, the coefficient of x^k in $f(x)g(x)$ is $a_0b_k + \cdots + a_kb_0$. These operations satisfy the basic rules of arithmetic, and the set of real polynomials is therefore an example of a ring (see Appendix C). In practice, if you have to multiply two polynomials by hand, simply use the distributive, associative and commutative laws.

If $f(x) = a_0 + a_1 x + \cdots + a_n x^n$, and r is a real number, then we define

$$f(r) = a_0 + a_1 r + \cdots + a_n r^n$$

and we say that r is a *root* of $f(x)$ if $f(r)=0$. You learned in high school that a polynomial of degree n has at most n distinct roots. If $f(x)$ and $g(x)$ are polynomials, then one can verify by calculation that $(f+g)(r)=f(r)+g(r)$ and $fg(r)=f(r)g(r)$.

In proving the various facts quoted above, the only properties of the real numbers that are used is the fact that they can be added, subtracted, multiplied and divided, and that the various standard rules of arithmetic hold. Recall from Chapter 2 (and Appendix C) that these properties can be summarized by saying that the set of real numbers is a *field*. But, as a result of our investigations in Chapter 2, we now know of another field, namely the set $\mathbb{Z}_p$, where p is a prime. It therefore seems appropriate to consider the set of polynomials with coefficients from $\mathbb{Z}_p$. This set is denoted $\mathbb{Z}_p[x]$. A typical such polynomial, for example, might be (take $p=5$) $[2]+[1]x+[4]x^2$. This looks ungainly, which is why we adopted the convention mentioned in the first paragraph above to drop brackets. With this convention in place, we can write this polynomial as $2+x+4x^2$. This looks nicer, but we have to keep in mind that all calculations are taking place mod p. For example, if we denote this polynomial $f(x)$, then $f(0)=2$, $f(1)=2$, $f(2)=0$, $f(3)=1$, and $f(4)=0$. Thus, this polynomial has two roots.

Polynomials with coefficients in $\mathbb{Z}_p$ behave in many respects like polynomials with real coefficients. In particular, we have the following theorem, which we will need in the next section.

Theorem 5.3.1

If $f(x)$ is a nonzero polynomial of degree n in $\mathbb{Z}_p[x]$, then $f(x)$ has at most n roots in $\mathbb{Z}_p$.

We will prove this theorem shortly, but first we note that if we were to consider polynomials in $\mathbb{Z}_n[x]$, where n is not a prime, this statement is not necessarily true. For example, the polynomial $3x^2$ in $\mathbb{Z}_{12}[x]$ has degree 2, but one can easily check (do so!) that it has 5 roots: 0, 2, 4, 8 and 10.

We now finish this section by giving a proof of Theorem 5.3.1.

Proof of theorem 5.3.1. The proof will be by Strong Induction on the degree n of the polynomial $f(x)$. If $n=0$ then $f(x)$ is a nonzero constant polynomial, which obviously has no roots, so the result is true in that case. Next, assume the result is true for all polynomials of degree less than n, and let us prove it is true for a polynomial $f(x)$ of degree n. Write $f(x) = a_0 + a_1 x + \cdots + a_n x^n$, and assume, hoping for a contradiction, that $f(x)$ has $n+1$ roots $r_1, \ldots, r_{n+1}$. Let $g(x)$ be the polynomial $a_n(x-r_1)\ldots(x-r_n)$. Observe that the highest degree term of

$g(x)$ is $a_n x^n$. Thus, the degree of the polynomial $k(x)=f(x)-g(x)$ is strictly less than n, and our Strong Induction hypothesis applies to it: this polynomial, if nonzero, cannot have n roots.

We also note, however, that $k(x)$ *does* have n roots: clearly, each r_i ($i=1, 2, \ldots, n$) is a root. So, by induction hypothesis, $k(x)$ must be the zero polynomial, which implies that $f(x)=g(x)$. But this in turn implies that $0=f(r_{n+1})=g(r_{n+1})=a_n(r_{n+1}-r_1)\cdots(r_{n+1}-r_n)$. But this is impossible: the product on the right-hand side of this equation is a product of nonzero elements of $\mathbb{Z}_p$, and we know from Chapter 2 that the product of nonzero elements of $\mathbb{Z}_p$ must be nonzero. So, we have found the contradiction that we sought, and this finishes the induction argument.

As a special case of this theorem, we record, for use in the next section, the following corollary.

Corollary 5.3.2

The polynomial x^n-1 has at most n roots in $\mathbb{Z}_p$.

Exercises

5.11   Find all roots of x^2+7x+1 in $\mathbb{Z}_{11}$.

5.12   Prove that in $\mathbb{Z}_p[x]$ there is always a nonconstant polynomial with no roots at all.

5.13   Find all "square roots of -1" in $\mathbb{Z}_{13}$.

5.14   Suppose you were to attempt to find all roots of x^2+4x+1 in $\mathbb{Z}_{13}$ by using an analog of the "quadratic formula". Does your method work? (The previous exercise should be helpful.)

5.4 Primitive Roots Modulo a Prime

In this section, we prove that, given any prime p, there exists a primitive root mod p. Throughout this section, the letter p denotes a prime. We begin by proving a result about the Euler phi function.

Theorem 5.4.1

Let n be a positive integer with distinct positive divisors $d_1, \ldots, d_k$. Then

$$n = \varphi(d_1)+\cdots+\varphi(d_k)$$

Proof. We begin by defining a function F whose domain is the set of positive integers. Specifically, if n is a positive integer with distinct positive divisors $d_1, \ldots, d_k$, then define

$F(n) = \varphi(d_1) + \cdots + \varphi(d_k)$. What we want to show is that F is the identity function: $F(n) = n$. What we already know (see Exercise 4.6) is that F is multiplicative, as is, of course, the identity function. Since two multiplicative functions are equal if they agree on prime powers (why?), it is therefore sufficient to show $F(n) = n$ if n is a prime power.

So, let $n = p^m$. Observe that the positive divisors of n are $1, p, \ldots, p^m$. Thus, $F(n) =$

$\varphi(1) + \varphi(p) + \cdots + \varphi(p^m) = 1 + (p-1) + \cdots + (p^m - p^{m-1})$. Notice that the sum on the right-hand side of this equation is a telescoping one: every term, except p^m, is cancelled out by a term in the next summand. So $F(n) = p^m = n$, as desired. This completes the proof.

We have now developed enough machinery to prove the existence of a primitive root modulo a prime p. As we did previously, we will abuse notation by omitting brackets on elements of $\mathbb{Z}_p^*$ and will write a instead of $[a]$.

Theorem 5.4.2

If p is a prime integer, then there is a primitive root mod p.

Proof. Suppose not. Then the order of every integer that is not a multiple of p is strictly less than $p-1$. For every proper divisor d of $p-1$, let us count the number of elements in $\mathbb{Z}_p^*$ of order d. First observe that if a is such an element, then $a, a^2, \ldots, a^{d-1}$ are all distinct. On the other hand, each of these elements is also a root of the polynomial $x^d - 1$. (This is because $(a^m)^d = (a^d)^m = 1$.) By Corollary 4.3.2, these must be *all* the roots of this polynomial. Thus, any element of order d must be one of these elements. Of course, not every one of these elements has order d; in fact, we know that a^m has order (exactly) d if and only if m and d are relatively prime. Thus, the number of elements of order d is either 0 or $\varphi(d)$; in particular, this number is less than or equal to $\varphi(d)$.

Now, under the assumption that there are no primitive roots mod p, it follows that the order of every element in $\mathbb{Z}_p^*$ (there are $p-1$ of them) is a proper divisor of $p-1$. If we list these proper divisors as $d_1, \ldots, d_k$, then it follows that

$p - 1 = $ (number of elements of order d_1) $+ \ldots.$ (number of elements of order d_k)

$$\leq \varphi(d_1) + \ldots + \varphi(d_k)$$

$$< \varphi(d_1) + \ldots + \varphi(d_k) + \varphi(p-1)$$

$$= p - 1 \text{ (by Theorem 5.4.1)}.$$

Reading from left to right, this gives $p-1 < p-1$, an obvious contradiction. This contradiction proves the theorem.

An examination of the preceding proof shows that it proves more than the existence of primitive roots: it establishes that for any prime p, there are exactly $\varphi(d)$ elements of order d in $\mathbb{Z}_p^*$ for every divisor d of $p-1$. Another way to phrase this is to say that of the integers from 1 to $p-1$, exactly $\varphi(d)$ of them have order d mod p. In particular, there are $\varphi(p-1)$ primitive roots.

Exercises

5.15 If a is a primitive root mod the odd prime p, prove that $a^{(p-1)/2} \equiv -1 (\mathrm{mod}\ p)$.

5.16 Use Theorem 4.4.2 and the previous exercise to give another proof of Wilson's Theorem (Theorem 2.4.3).

5.17 If p is an odd prime and $a^{(p-1)/2} \equiv -1 (\mathrm{mod}\ p)$, is a necessarily a primitive root mod p? Explain.

5.18 Prove that any prime greater than 3 has an even number of primitive roots.

5.19 What primes p have $p-1$ as a primitive root? Justify your answer.

5.5 An Application: Diffie-Hellman Key Exchange

In this section, we revisit the subject of cryptography and give an application of the existence of primitive roots modulo a prime. Recall from our work in cryptography that some cryptosystems rely on the existence of a key—a number that is used in the implementation of the system. For example, a simple shift cypher has, as a key, the number of spaces that are shifted. If an eavesdropper discovers the key, then the cryptosystem is useless. So, if Alice wants to send Bob an encoded message, both Bob and Alice need to know the key, but transmitting this information via a possibly insecure channel poses risks. The question is whether Alice and Bob can both obtain this key in a way that protects this information from eavesdroppers. The Diffie-Hellman Key Exchange provides such a way.

The procedure can be described as follows. Alice and Bob agree on a (very large) prime p, and a primitive root g modulo p. The numbers p and g need not be kept secret. Alice then selects a secret number a between 1 and $p-1$, and Bob selects a secret number b between 1 and $p-1$. Alice then transmits the number g^a to Bob, and Bob transmits the number g^b to Alice. It does not matter if either of the numbers g^a or g^b are intercepted; there is no known computationally feasible way of determining a and b from knowledge of these numbers.

Now that Alice has the number g^b, she can compute $(g^b)^a$ (modulo p). Likewise, Bob can compute the number $(g^a)^b$ (modulo p). But of course these are the same number. So, Alice and Bob are now in possession of a shared number, which they can use as the key.

Let us illustrate this with an example, where, as we have done previously, we have made the numbers absurdly small. In particular, take $p=11$. We have seen previously that 2 is a primitive root mod 11, so let us take $g=2$. Suppose Alice selects $a=3$ and Bob selects $b=5$. Thus, Alice transmits the number 8 to Bob and Bob transmits the number 10 (i.e., 32 mod 11) to Alice. Alice then takes the number 10 and raises it to the 3rd power, getting 10. Bob takes the number 8 and raises it to the 5th power; a short calculation shows that this number, mod 11, is also 10 (as it had to be!). So, Alice and Bob now have a shared key.

Note an interesting thing: the Diffie-Hellman Method not only shares a key, it *creates* one. Neither Bob nor Alice knew what the key would be until the other one transmitted his or her number.

One final comment: an astute reader might ask at this point why it is necessary to select g to be a primitive root; the method would work no matter what the number g is. The reason we select g to be a primitive root is more practical than mathematical. The more distinct powers of g there are, the harder it is to find the exponent, knowing the power of g. And we know that primitive roots give the most distinct powers of g.

5.6 Another Application: ElGamal Cryptosystem

The Diffie-Hellman Key Exchange is not a cryptosystem; it does not tell you how to encrypt or decrypt a message. However, there is a cryptosystem that is reminiscent of Diffie-Hellman; in particular, it, like Diffie-Hellman, uses a primitive root g modulo a prime p. We discuss this method, the ElGamal system, in this section. As usual, we have two people, Alice and Bob; Bob wants to send a message secretly to Alice. We can assume that this message is a number, which we denote m.

The method works as follows. Alice selects a prime p and a primitive root g modulo p. She also selects an integer a between 2 and $p-2$. With this information, she can compute $g^a \bmod p$. Alice then sends p, g and g^a (mod p) to Bob; she doesn't care if the channel is insecure, as these numbers need not be secret. The number a, however, *is* a secret, and she doesn't transmit it to anybody.

Bob, in turn, selects an integer b between 2 and $p-2$. Having received g^a, he then computes (mod p) the number g^b and transmits it to Alice. He also transmits the number $g^{ab}m$ to her. Alice, having received g^b, can, as in Diffie-Hellman, calculate g^{ab} and can then compute its inverse mod p. Having done this, she can multiply it by $g^{ab}m$ to recover the secret message m.

Let us illustrate the method by using the same (admittedly unrealistically small) numbers that we used in our example of Diffie-Hellman in the previous section: $p=11$, $g=2$, $a=3$ and $b=5$. Suppose Bob wants to transmit the message $m=4$ to Alice. So, Alice transmits the numbers 11, 2 and 8 to Bob. Bob then computes $g^b \bmod 11$; as we saw in the last section, this is 10. Multiplying this by $m=4$ gives (mod 11) the number 7. So Bob transmits to Alice the numbers 10 and 7. The multiplicative inverse of 10 mod 11 is itself, so Alice multiplies 10 by 7, getting 70, which is congruent to 4 (=m). Thus, Alice has recovered the secret message m.

Challenge Problems for Chapter 5

C5.1   If $n>1$ is an integer, prove that n does not divide 2^n-1. (Hint: suppose not, and let p be the smallest prime factor of n. Then consider the order of 2 mod p.)

C5.2   If $n>1$ is an integer, prove that n does divide $\varphi(2^n-1)$.

C5.3   Prove that if m and n are relatively prime integers, each greater than 1, then mn does not have a primitive root.

C5.4   Prove that if a and b have orders r and s, respectively, modulo n, and r and s are relatively prime, then ab has order rs mod n.

C5.5   If the integer a has order 3 modulo the prime p, prove that $1+a+a^2$ is divisible by p.

C5.6   Under the circumstances of the previous problem, prove that a^2 has order 6 mod p.

6

Quadratic Reciprocity

In this chapter, we consider integers that are congruent, modulo a prime p, to the square of another integer. This is equivalent to considering when an element of $\mathbb{Z}_p$ can be written as the square of another element of $\mathbb{Z}_p$. For example, in $\mathbb{Z}_{11}$, [3] is a square, because $[3]=[5]^2$. Our ultimate objective is a discussion of one of the most famous theorems of elementary number theory, the Law of Quadratic Reciprocity.

Throughout this chapter, the letter p denotes an arbitrary but fixed odd prime integer. We will also frequently continue the practice of omitting brackets when discussing elements of $\mathbb{Z}_p$.

6.1 Squares Modulo a Prime

We begin with an important definition.

Definition 6.1.1

Let p be an odd prime. We say that the integer a (not a multiple of p) is a *quadratic residue mod p* if the equation $x^2 \equiv a \pmod{p}$ has an integer solution. If the equation does not have a solution, then a is a *quadratic nonresidue mod p*. Equivalently, a is a quadratic residue mod p if and only if, for some integer x, $[a]=[x]^2$ in $\mathbb{Z}_p$.

A trivial consequence of this definition is that if a and b are two integers that are congruent to each other modulo p, and if one of these integers is a quadratic residue mod p, then so is the other. We leave the proof of this simple result to the exercises.

As an example, let us determine the quadratic residues mod 7. The simplest way to do this is to just square each of the nonzero elements of $\mathbb{Z}_7$: we get, after a simple calculation, [1], [4] and [2]. So the quadratic residues mod 7 are the integers that are congruent to 1, 2 or 4 mod 7. The quadratic nonresidues mod 7 are the integers congruent to 3, 5 and 6.

When doing the calculation above, we note that there is duplication as we square things: $[1]^2=[-1]^2=[6]^2$, etc. In general, assuming that p is an odd prime, if we list the elements of $\mathbb{Z}_p^*$ from [1] to $[p-1]$ and start to square them, we can pair off the first and last terms, the second and next-to-last, and so

DOI: 10.1201/9781003318712-7

forth. Since we are assuming that p is an odd prime, there is no duplication as we square things and we arrive at a total of $(p-1)/2$ different squares. This simple idea provides the method of proving the following theorem.

Theorem 6.1.1

If p is an odd prime, then there are exactly $(p-1)/2$ quadratic residues that are non-congruent mod p.

Proof. We employ the reasoning above. As noted there, if a is a quadratic residue mod p then $[a]$ must be one of $[1]^2, \ldots, [(p-1)/2]^2$. Thus, there are at most $(p-1)/2$ choices for a. If we knew that $[1]^2, \ldots, [(p-1)/2]^2$ were all *distinct* elements in $\mathbb{Z}_p^*$, then we would know that there are *exactly* $(p-1)/2$ choices for a, and we would be done.

So, let us prove that. Assume to the contrary that there is some duplication; let us say $[a]^2 = [b]^2$ where, without loss of generality, we have $0 < a < b \le (p-1)/2$. This means that p divides $b^2 - a^2 = (b-a)(b+a)$. By Euclid's Lemma, we then know that p must divide either $b-a$ or $b+a$. But both of these terms are positive integers that are strictly less than p, so this is impossible. This contradiction proves the result.

Exercises

6.1   Find all quadratic residues mod 17.

6.2   Find all quadratic residues mod 23.

6.3   Prove that the product of two quadratic residues mod p is a quadratic residue mod p.

6.4   Prove that the product of a quadratic residue and a quadratic non-residue mod p is a quadratic nonresidue mod p.

6.5   Can a primitive root mod p (an odd prime) be a quadratic residue mod p? Explain.

6.6   Find the smallest odd prime p for which -1 is a quadratic residue mod p.

6.2 Euler's Criterion and Legendre Symbols

Let us begin this section by posing a problem: if p is an odd prime and a is an integer that is relatively prime to p, what is $a^{(p-1)/2}$ congruent to mod p? Of course, the best way to tackle a problem like this is to look at some specific cases and see if we can discern a pattern, so let us take $p=7$. In this case $(p-1)/2=3$, so we take cubes of integers that are relatively prime to 7. Some calculation shows that 1^3, 2^3 and 4^3 are all congruent to 1 mod 7, and 3^3, 5^3 and 6^3 are all congruent to -1 mod 7. This seems fairly random, until we compare with the example that led off Section 6.1. We noticed there that 1, 2 and 4 were quadratic residues mod 7 and 3, 5 and 6 were quadratic nonresidues. We now have a pattern and a conjecture: $a^{(p-1)/2}$ is congruent to 1 mod p if a is a quadratic residue mod p, and is congruent to -1 otherwise. This conjecture is in fact true, and the statement of the resulting theorem is often called Euler's Criterion. The result, which we will now prove, can provide a useful way of determining whether a given integer is, or is not, a quadratic residue mod p. The proof is a nice application of the existence of a primitive root modulo a prime.

Theorem 6.2.1

(Euler's Criterion) If p is an odd prime and a is an integer that is relatively prime to p, then $a^{(p-1)/2}$ is congruent to 1 mod p if a is a quadratic residue mod p, and is congruent to -1 otherwise.

Proof. We first show that $a^{(p-1)/2}$ is congruent to either 1 or -1 mod p. If we denote this integer by x, then we have $x^2 \equiv a^{p-1} \pmod{p} \equiv 1 \pmod{p}$, by Fermat's Little Theorem. So p divides $x^2 - 1 = (x-1)(x+1)$, and by Euclid's Lemma, this means that p divides either $x - 1$ or $x+1$. So x is congruent to either 1 or -1 mod p, as claimed.

Next, suppose that a is a quadratic residue mod p. Then we can write $x^2 \equiv a \pmod{p}$ for some integer x. Raising both sides to the $((p-1)/2)^{\text{th}}$ power and using Fermat's Little Theorem, we see that $a^{(p-1)/2} \equiv x^{p-1} \pmod{p}$

$$\equiv 1 \pmod{p}, \text{ as desired.}$$

Finally, assume that a is a quadratic nonresidue mod p. We want to show that $a^{(p-1)/2}$ is congruent to -1 mod p. Assume, hoping for a contradiction, that $a^{(p-1)/2} \equiv 1 \pmod{p}$. Let g be a primitive root mod p. Then $a \equiv g^m \pmod{p}$ for some integer m. It follows that

$$1 \equiv a^{(p-1)/2} \pmod{p}$$

$$\equiv g^{m(p-1)/2} \pmod{p}.$$

Because the order of g mod p is $p - 1$, it follows from the above that $m(p-1)/2$ must be a multiple of $(p-1)/2$; i.e., $m/2$ must be an integer. Thus, $m = 2k$ for some integer k. But then

$$a \equiv g^m \pmod{p}$$

$$\equiv g^{2k} \pmod p$$

$$\equiv (g^k)^2 \pmod p.$$

But this congruence equation says that a is a quadratic residue mod p, a contradiction. This contradiction proves the theorem.

This theorem tells us that, if a is not a multiple of the odd prime p, the number $a^{(p-1)/2}$, mod p, is a quantity that is either 1 or –1 depending on whether a is, or is not, a quadratic residue mod p. This suggests it would be helpful if we had a compact symbol to represent this. This leads to our next definition.

Definition 6.2.2

If p is an odd prime and a is an integer relatively prime to p, then the *Legendre symbol* $\left(\dfrac{a}{p}\right)$ is defined to be 1 if a is a quadratic residue mod p and –1 if a is a quadratic nonresidue mod p.

So, for example, $\left(\dfrac{2}{11}\right)=-1$, and $\left(\dfrac{5}{11}\right)=1$, as the reader may verify. It is also obvious that $\left(\dfrac{1}{p}\right)=1$ for all p, and that if $a \equiv b \pmod p$, then $\left(\dfrac{a}{p}\right)=\left(\dfrac{b}{p}\right)$. Note also the following result, which is really nothing more than a restatement of Euler's Criterion.

Theorem 6.2.3

If p is an odd prime and a is an integer that is relatively prime to p, then $a^{(p-1)/2} \equiv \left(\dfrac{a}{p}\right) \pmod p$.

Proof. If a is a quadratic residue mod p, both sides of this congruence are congruent to 1, and if a is not a quadratic residue, both sides are congruent to –1.

We can use this result to establish a "multiplicative property" of the Legendre symbol.

Theorem 6.2.4

If p is an odd prime and a and b are integers that are relatively prime to p, then $\left(\dfrac{ab}{p}\right)=\left(\dfrac{a}{p}\right)\left(\dfrac{b}{p}\right)$.

Proof. Since 1 is not congruent to $-1 \bmod p$, it suffices, to show that $\left(\dfrac{ab}{p}\right) = \left(\dfrac{a}{p}\right)\left(\dfrac{b}{p}\right)$, to show that both sides are congruent mod p. This, in turn, is an immediate consequence of the previous theorem: $\left(\dfrac{ab}{p}\right) \equiv (ab)^{(p-1)/2} \equiv a^{(p-1)/2}b^{(p-1)/2} \equiv \left(\dfrac{a}{p}\right)\left(\dfrac{b}{p}\right)$, with, of course, all congruences being mod p.

It follows from this that the product of two quadratic residues or two quadratic nonresidues is a quadratic residue, and the product of a quadratic residue and a quadratic nonresidue is a quadratic nonresidue. Some of these results can be easily proved directly, and appeared as Exercises 6.3 and 6.4 in the last section.

As an application of these ideas, we can answer the question: *for which odd primes p is -1 a quadratic residue mod p?* We know that $\left(\dfrac{-1}{p}\right) \equiv (-1)^{(p-1)/2} \pmod{p}$.

For the right-hand side to be equal to 1, it must be the case that $(p-1)/2$ is even, say $(p-1)/2 = 2k$ for some integer k. But this happens if and only if $p = 4k+1$. We have thus proved:

Theorem 6.2.5

If p is an odd prime, then -1 is a quadratic residue mod p if and only if $p \equiv 1 \pmod 4$.

As a nice application of this result, we can prove a theorem that generalizes the result, proved much earlier in the book, that there are infinitely many primes.

Theorem 6.2.6

There are infinitely many primes of the form $4n+1$, for a positive integer n.

Proof. Suppose to the contrary (hoping for a contradiction) that there only a finite number of such primes, and let us denote the product of all of them by P. Now consider the number $N = (2P)^2 + 1$. Clearly, N is an odd integer greater than 1, so it has a prime factor p, which must be odd since N is. Clearly also, p does not divide P and so cannot be one of the primes whose product defined P. In other words, p cannot be one of the finite number of primes existing of the form $4n+1$. Note, however, that since p divides N, we must have $(2P)^2 \equiv -1 \pmod{p}$, an equation which says that -1 is a quadratic residue mod p. But if this is the case, then Theorem 6.2.5 tells us that p *is* of the form $4n+1$, a contradiction.

A careful reader will recall that we pointed out, much earlier (Theorem 1.5.9) that there is a far-reaching generalization of this result known as Dirichlet's Theorem, the proof of which is beyond the scope of the text: if a and b are two relatively prime positive integers, then there are infinitely many primes of the form $an+b$.

Exercises

6.7 Evaluate $\left(\dfrac{8}{11}\right)$, $\left(\dfrac{12}{11}\right)$ and $\left(\dfrac{-2}{11}\right)$.

6.8 If both a and $-a$ are quadratic residues mod p, what can you say about p?

6.9 Evaluate $\left(\dfrac{97}{101}\right)$. (Do this mentally.)

6.10 Let q be the smallest positive nonresidue mod the odd prime p. Prove that q is prime.

6.11 If p is an odd prime, evaluate, with proof, $\left(\dfrac{1}{p}\right)+\left(\dfrac{2}{p}\right)+...+\left(\dfrac{p-1}{p}\right)$.

6.3 The Law of Quadratic Reciprocity

If p and q are distinct odd primes, then two apparently distinct and unrelated questions arise: is p a quadratic residue mod q? Is q a quadratic residue mod p? There is no reason to think, *a priori*, that the answer to one question should affect the answer to the other. In fact, however, the answer to one question *completely determines* the answer to the other. The precise statement of this remarkable fact is known as the *Law of Quadratic Reciprocity*, which we will discuss in this section.

The Law of Quadratic Reciprocity historically arose in connection with the attempt to discuss what primes could be expressed in the form ax^2+by^2 for certain constants a and b. For example, taking $a=b=1$, the question would be: what primes can be expressed as a sum of two squares? Euler, in the 18th century, realized that these questions led to the questions posed in the paragraph above. He stated, but did not succeed in proving, the Law of Quadratic Reciprocity that is stated below. The first person to actually record a proof of the result was Gauss, in 1796. Gauss was fascinated by this result and returned to it frequently over the course of his life, eventually supplying eight different proofs (six of them published during his life, two found in his papers after his death). There are now more than 200 known proofs of the result, making it perhaps second only to the Pythagorean Theorem in the number of proofs known for it.

With this as (hopefully) dramatic buildup, we now state the result.

Theorem 6.3.1 (Law of Quadratic Reciprocity.)

If p and q are distinct odd primes, then

$$\left(\frac{p}{q}\right)\left(\frac{q}{p}\right) = (-1)^{\frac{1}{2}(p-1)\frac{1}{2}(q-1)}.$$

Let's take a second to think about what this means. The term on the right is known to us if we know p and q; it is either 1 or –1, depending on whether the exponent is even or odd. If it is even and the right hand side is 1, that means the two Legendre symbols on the left are either both 1 or both –1; i.e., p is a quadratic residue mod q if and only if q is a quadratic residue mod p. If, on the other hand, the right-hand side is –1, then that means the two Legendre symbols have opposite signs, which means that p is a quadratic residue mod q if and only if q is *not* a quadratic residue mod p.

Any odd number is, of course, congruent to either 1 or 3 mod 4. It is easy to show (we leave this as an exercise) that the exponent on the right-hand side above is even if either p or q is congruent to 1 mod 4 and is odd if both are congruent to 3 mod 4. Thus, we may rephrase Theorem 6.3.1 above as follows: If p and q are odd primes, both congruent to 3 mod 4, then p is a quadratic residue mod q if and only if q is *not* a quadratic residue mod p; if, on the other hand, either p or q is congruent to 1 mod 4, then p is a quadratic residue mod q if and only if q *is* a quadratic residue mod p.

Although the statement of this law is elegant and beautiful, the same cannot be said for its elementary proofs. There are, as previously noted, a lot of known proofs of this result, but all of the ones that are elementary enough to be presented in a first course in number theory are fairly technical counting arguments that seem to miraculously turn out right at the end. None of them, unfortunately, give any real feeling for *why* the result should be true. For that reason, we will omit the proof. Proofs are easily found in other elementary number theory textbooks, such as [KW], as well as in journals, such as Kim's relatively recent proof [Kim].

We mention at this point that the result we have stated in Theorem 6.3.1 is not the entire Law of Quadratic Reciprocity. There are also two "supplemental relations", which we will state and prove in the next section. (One of them, in fact, has already been proved.) For the moment, however, we focus on this "main" result.

Let us illustrate the usefulness of the result in determining whether a prime p is or is not a quadratic residue modulo another prime q. Consider, for example, $p=3$ and $q=101$. It would be, to put it mildly, a tedious chore to square the integers from 1 to 50 to determine whether any of those squares are congruent to 3 mod 101. But with the law of quadratic reciprocity, the calculation becomes so trivial that it could be done mentally. Since 101 is congruent to 1 mod 4, we know that $\left(\dfrac{3}{101}\right)=\left(\dfrac{101}{3}\right)$. But since 101 is congruent to 2 mod 3,

$\left(\dfrac{101}{3}\right)=\left(\dfrac{2}{3}\right)=-1$. So, 3 is not a quadratic residue mod 101.

As another example, we ask whether 97 (which is a prime) is a quadratic residue mod 101. Just as before, $\left(\dfrac{97}{101}\right)=\left(\dfrac{101}{97}\right)=\left(\dfrac{4}{101}\right)$, where the last equality derives from the fact that 101 is congruent to 4 mod 97. But 4, being a perfect square, is obviously a quadratic residue mod 101, so $\left(\dfrac{97}{101}\right)=\left(\dfrac{4}{101}\right)=1$. So we have shown, with minimal calculation, that 97 is a quadratic residue mod 101.

We can also use the Law of Quadratic Reciprocity to investigate the quadratic character of non-primes modulo a prime. For example, suppose we want to determine whether 57 is a quadratic residue mod 101. Now, 57 is not a prime; its prime factors are 3 and 19. However, using Theorem 6.2.4, we see that $\left(\dfrac{57}{101}\right)=\left(\dfrac{3}{101}\right)\left(\dfrac{19}{101}\right)$. So, it suffices to evaluate each of the Legendre symbols on the right. We have already determined that $\left(\dfrac{3}{101}\right)=-1$, and we see that $\left(\dfrac{19}{101}\right)=\left(\dfrac{101}{19}\right)=\left(\dfrac{6}{19}\right)=\left(\dfrac{2}{19}\right)\left(\dfrac{3}{19}\right)$. It is not too hard to verify that 2 is not a quadratic residue mod 19. It is equally easy to see that 3 is not either, but we can do this without calculation (since 3 and 19 are both congruent to 3 mod 4) by noting that $\left(\dfrac{3}{19}\right)=-\left(\dfrac{19}{3}\right)=-\left(\dfrac{1}{3}\right)=-1$. Thus, putting everything together, we see that $\left(\dfrac{57}{101}\right)=\left(\dfrac{3}{101}\right)\left(\dfrac{19}{101}\right)=\left(\dfrac{3}{101}\right)\left(\dfrac{2}{19}\right)\left(\dfrac{3}{19}\right)=(-1)^3=-1$. In other words, 57 is not a quadratic residue mod 101.

Exercises

6.12 Evaluate $\left(\dfrac{17}{101}\right),\left(\dfrac{13}{101}\right),\left(\dfrac{51}{101}\right)$ and $\left(\dfrac{11}{53}\right)$. Explain your answers.

6.13 Evaluate $\left(\dfrac{3}{19}\right)$ in two ways, using Euler's Criterion and Quadratic Reciprocity.

6.14 Prove the "rephrased" version of Theorem 6.3.1 that is stated in the text.

6.15 Find (with proof) all odd primes p for which 5 is a quadratic residue mod p.

6.16 If p is an odd prime, prove that 3 is a quadratic residue mod p if and only if p is congruent to 1 or 11 mod 12.

6.4 The Supplemental Relations

We noted in the last section that there are two "supplemental relations" to the main Law of Quadratic Reciprocity, one of which we have already seen (as Theorem 6.2.5.). We state them both below, prove the second one and then look at some illustrations of its use.

Supplemental Relation 1: If p is an odd prime, then -1 is a quadratic residue mod p if and only if $p \equiv 1 \pmod 4$.

Supplemental Relation 2: If p is an odd prime, then 2 is a quadratic residue mod p if and only if p is congruent to either 1 or 7 (mod 8).

As noted above, the first of these supplemental relations has already been stated and proved, so we turn to the second. We shall use Euler's Criterion to prove the result. This directs us to consider the number $2^{(p-1)/2}$; we must show this number is congruent to 1 mod p if p is congruent to 1 or 7 mod 8 and is congruent to -1 mod p if p is congruent to 3 or 5 mod 8.

We first assume that that p is congruent to 3 mod 8. To illustrate the method of proof, we do a simple calculation with $p=11$. Of course, determining what 2^5 is congruent to mod 11 is child's play and can be done mentally, but we will do the computation in such a way as to make clear how this argument generalizes. In the calculations that follow, all congruences are, of course, mod 11.

The trick is to compute $2^5(5!)$ rather than 2^5. We have

$$2^5(5!) = 2^5(1)(2)(3)(4)(5)$$

$$= (2 \times 1)\,(2 \times 2)\,(2 \times 3)\,(2 \times 4)\,(2 \times 5)$$

$$\equiv (2)(4)(-5)(-3)\,(-1)$$

$$\equiv (-1)^3\,(5!)$$

$$\equiv (-1)5!$$

Cancellation (which is permissible, because 5! is relatively prime to 11) then shows that $2^5 \equiv -1 \pmod{11}$, as desired.

To deal with the case of a general prime p that is congruent to 3 mod 8, write $p=8k+3$ for some integer k. Then of course $(p-1)/2=4k+1$. Euler's Criterion requires consideration of $2^{(p-1)/2}\,((p-1)/2)! =$

$$2^{4k+1}(4k+1)! = 2^{4k+1}(1)(2)\ldots(4k+1)$$

$$= (2)(4)\ldots(8k+2)$$

which is congruent, mod $p=8k+3$, to

$$(2)(4)\ldots(4k)(-(4k+1))\ldots(-1) \tag{*}$$

Let us look at this last product closely. The first $2k$ terms are all the even positive integers that are less than or equal to $4k+1$. The remaining $2k+1$ terms are the negatives of all the odd integers that are less than or equal to $4k+1$. If we factor out a minus sign from these last $2k+1$ terms, the product (*) is seen to be equal to $(-1)^{4k+1}$ times the product of all positive even integers $\leq 4k+1$, times the product of all positive odd integers $\leq 4k+1$. But these last two products, multiplied together, is just $(4k+1)!$ Since it is obvious that $(-1)^{4k+1}=-1$, we have just shown that $2^{(p-1)/2}((p-1)/2)!$ is congruent mod p to $(-1)^{((p-1)/2)!}$, which, upon cancellation, shows that $2^{(p-1)/2}$ is congruent to -1 mod p. In other words, by Euler's Criterion, if p is congruent to 3 mod 8, then 2 is a quadratic nonresidue mod p.

Similar calculations can be used to verify the Second Supplemental Relation in the cases where p is congruent mod 8 to 1, 5 or 7. We leave these calculations as an exercise. We end this section by giving some examples illustrating how these Supplemental Relations can be applied.

For example, suppose we want to know whether 32 is a quadratic residue mod 101. Since $32=2^5$, the multiplicative property of the Legendre symbol gives $\left(\dfrac{32}{101}\right)=\left(\dfrac{2}{101}\right)^5$. Since 101 is congruent to 5 mod 8, this is $(-1)^5=-1$. So the answer is no, 32 is not a quadratic residue mod 101.

As another example, let us explore the quadratic character of 12 modulo the prime 43. Again, we use the multiplicative property of the Legendre symbol to conclude that $\left(\dfrac{12}{43}\right)=\left(\dfrac{2}{43}\right)\left(\dfrac{6}{43}\right)$. We clearly have $\left(\dfrac{2}{43}\right)=-1$ because 43 is congruent to 3 mod 8. As for $\left(\dfrac{6}{43}\right)$, the slickest way to deal with this is to notice that, modulo 43, 6 is congruent to the perfect square 49, and so is a quadratic residue: $\left(\dfrac{6}{43}\right)=1$. Thus, $\left(\dfrac{12}{43}\right)=\left(\dfrac{2}{43}\right)\left(\dfrac{6}{43}\right)=-1$, and 12 is not a quadratic residue mod 43.

Exercises

6.17 If the prime p is congruent to 1 mod 8, determine whether $p-2$ is a quadratic residue mod p. Prove your answer.

6.18 Evaluate $\left(\dfrac{18}{37}\right)$ both using, and not using, the Second Supplemental Relation.

6.19 Evaluate $\left(\dfrac{30}{37}\right)$.

6.5 The Jacobi Symbol

The Legendre symbol $\left(\dfrac{a}{p}\right)$ was defined only for odd primes p in the lower position. In this section of the text, we show how the Legendre symbol can be generalized for any odd positive integer n in the lower position. The resulting symbol $\left(\dfrac{a}{n}\right)$ is called the Jacobi symbol.

Definition 6.5.1

If $n > 1$ is an odd positive integer with prime factorization $n = p_1 \ldots p_m$ and a is a nonzero integer that is relatively prime to n, the *Jacobi symbol* $\left(\dfrac{a}{n}\right)$ is defined to be the product of Legendre symbols $\left(\dfrac{a}{p_1}\right) \ldots \left(\dfrac{a}{p_m}\right)$. The Jacobi symbol $\left(\dfrac{a}{1}\right)$ is defined to be 1.

The primes that appear in the factorization of n above need not, of course, be distinct. Because the prime factorization of a positive integer greater than 1 is unique, this definition is well-defined. Of course, if n is itself prime, then the Jacobi symbol $\left(\dfrac{a}{n}\right)$ is the same thing as the Legendre symbol $\left(\dfrac{a}{n}\right)$.

As a simple example, let us compute the Jacobi symbol $\left(\dfrac{7}{33}\right)$. By definition, this is $\left(\dfrac{7}{3}\right)\left(\dfrac{7}{11}\right)$. The first term on the right is obviously 1 because 7 is congruent to 1 mod 3. The second term can be easily computed directly or can be computed using the Law of Quadratic Reciprocity: $\left(\dfrac{7}{11}\right) = -\left(\dfrac{11}{7}\right) = -\left(\dfrac{4}{7}\right) = -1$. So $\left(\dfrac{7}{33}\right) = -1$.

Any Jacobi symbol, as a product of Legendre symbols, must be either 1 or -1. If a Jacobi symbol $\left(\dfrac{a}{n}\right)$ is -1, then a is not a square mod n. (See the exercises.) However, if a Jacobi symbol $\left(\dfrac{a}{n}\right)$ is 1, that does not mean that a is a perfect square mod n. As an illustration, note that 5 is not a square mod 9 (any square mod 9 must be congruent to one of 1, 4 or 7), yet $\left(\dfrac{5}{9}\right) = 1$.

Jacobi symbols satisfy certain basic properties, collected in the theorem below. Because these properties all follow immediately from either the definition or the corresponding properties of Legendre symbols, the proof of the theorem is left as an exercise.

Theorem 6.5.2

Let m and n be positive odd integers and a and b nonzero integers that are relatively prime to both m and n. Then

(a) $\left(\dfrac{a}{n}\right)\left(\dfrac{b}{n}\right)=\left(\dfrac{ab}{n}\right)$

(b) $\left(\dfrac{a}{mn}\right)=\left(\dfrac{a}{m}\right)\left(\dfrac{a}{n}\right)$

(c) $\left(\dfrac{a}{n}\right)=1$ if n is a square

(d) If a and b are congruent mod n, then $\left(\dfrac{a}{n}\right)=\left(\dfrac{b}{n}\right)$

Jacobi symbols also satisfy a Quadratic Reciprocity Law (complete with supplemental relations), which we state below but do not prove.

Theorem 6.5.3

If m and n are positive odd relatively prime integers, then

(e) $\left(\dfrac{m}{n}\right)\left(\dfrac{n}{m}\right)=(-1)^{\frac{1}{2}(m-1)\frac{1}{2}(n-1)}$

(f) $\left(\dfrac{-1}{n}\right)$ is equal to 1 if $n \equiv 1$ (mod 4), and equal to 1 if $n \equiv 3$ (mod 4).

(g) $\left(\dfrac{2}{n}\right)$ is equal to 1 if $n \equiv 1, 7$ (mod 8), and equal to -1 if $n \equiv 3, 5$ (mod 8).

One advantage to using these identities for Jacobi symbols is that, in contrast with Legendre symbols, we don't have to concern ourselves with the question of whether the entries in the symbol are primes. For very large integers, it may be difficult to conveniently determine whether that integer is or is not a prime, or to factor a known composite integer into primes; if we think of $\left(\dfrac{m}{n}\right)$ as being a Jacobi symbol rather than a Legendre symbol, we no longer have to worry about this.

To illustrate these ideas and to show how Jacobi symbols can sometimes simplify the computation of Legendre symbols, consider the Legendre symbol $\left(\dfrac{105}{113}\right)$. Without using Jacobi symbols, we would have to factor 105 and then separately compute the three Legendre symbols $\left(\dfrac{3}{113}\right)$, $\left(\dfrac{5}{113}\right)$ and $\left(\dfrac{7}{113}\right)$. However, using Jacobi symbols, we can cheerfully ignore the fact that

105 is not a prime and apply Theorem 6.5.3 directly: $\left(\dfrac{105}{113}\right)=\left(\dfrac{113}{105}\right)=\left(\dfrac{8}{105}\right)=$ $\left(\dfrac{2}{113}\right)^3=1.$

As a final example, let us determine whether 55 is a quadratic residue modulo the odd prime 401. We can use Jacobi symbols to evaluate the Legendre symbol $\left(\dfrac{55}{401}\right)$ directly (and very quickly) rather than computing $\left(\dfrac{5}{401}\right)$ and $\left(\dfrac{11}{401}\right)$. By Theorem 6.5.3(a), $\left(\dfrac{55}{401}\right)=\left(\dfrac{401}{55}\right)=\left(\dfrac{16}{55}\right)$, which is obviously 1 because 16 is a square.

Exercises

6.20 Prove that if a Jacobi symbol $\left(\dfrac{a}{n}\right)$ is –1, then a is not a square mod n.

6.21 Evaluate $\left(\dfrac{105}{113}\right)$ without using Jacobi symbols.

6.22 Evaluate $\left(\dfrac{55}{401}\right)$ without using Jacobi symbols.

6.23 Prove Theorem 6.5.2.

6.24 Prove part (b) of Theorem 6.5.3.

6.25 Evaluate the Jacobi symbol $\left(\dfrac{109}{385}\right)$.

Challenge Problems for Chapter 6

C6.1 Suppose that q and $p=2q+1$ are both odd primes. Prove that the primitive roots of p consist of the quadratic nonresidues of p and one other number. What is that number?

C6.2 With p and q as in the previous problem, prove that -4 is a primitive root of p.

C6.3 Prove that every element of $\mathbb{Z}_p$ can be written as a sum of two squares of elements of $\mathbb{Z}_p$.

C6.4 If $p > 5$ is a prime, prove that there are two consecutive quadratic residues mod p.

(Hint: Show that at least one of the numbers 2, 5 and 10 is a quadratic residue mod p.)

C6.5 If $p > 5$ is a prime, prove that there are two quadratic residues mod p that differ by 2.

C6.6 Let $p=4k+1$ be a prime, and d an odd divisor of k. Prove that d is a quadratic residue mod p.

C6.7 Let $n \geq 4$. Prove that $N = 1! + 2! + \ldots + n!$ is never a square. (Hint: Consider N mod 5.)

C6.8 Prove that 3 is a quadratic nonresidue of any prime of the form $4^n + 1$.

7

Arithmetic Beyond the Integers

Up to now, we have been studying the set $\mathbb{Z}$ integers. In this chapter, however, we expand our horizons and study other number systems that have features in common with $\mathbb{Z}$, but also some differences as well. We study them for several reasons. First, we can use these systems to actually prove things *about* the integers, and second, their study helps shed some light on unique factorization into primes, which turns out to be a subtler idea than one might expect at first.

7.1 Gaussian Integers: Introduction and Basic Facts

The first new number system that we will study in this chapter is the set $\mathbb{Z}[i]$ of *Gaussian integers*. Useful references for this material include [Con] and [KW]. We begin with a review of complex numbers, which the reader presumably has seen before in high school.

A *complex number* is a number of the form $a + bi$, where a and b are real numbers and i is the so-called "imaginary unit" characterized by the equation $i^2 = -1$. On the set $\mathbb{C}$ of complex numbers, operations of addition, subtraction, multiplication and division (by nonzero complex numbers) are defined that satisfy all the usual rules of arithmetic—associative law, distributive law, etc. (For those familiar with abstract algebra, this means that $\mathbb{C}$, with these operations, is a *field*.) Addition and subtraction are defined "component-wise":

$$(a + bi) \pm (c + di) = (a + c) \pm (b + d)i$$

Multiplication can be given by a formula, but it's easier to just think of it as a consequence of the distributive and associative laws:

$$(a + bi)(c + di) = (a + bi)c + (a + bi)di$$

$$ac + bic + adi + (bd)i^2$$

$$ac + bic + adi - bd$$

$$(ac - bd) + (bc + ad)i$$

DOI: 10.1201/9781003318712-8

Note for future reference that, in particular, $(a+bi)(a-bi)=a^2+b^2$.

If z denotes the complex number $a+bi$, then the complex number $a - bi$ is called the complex conjugate of z and is denoted $\bar{z}$. A standard fact, easily verified by direct computation, is that conjugation is multiplicative: if z and w are complex numbers, then $\overline{zw} = \bar{z} \cdot \bar{w}$.

Likewise, rather than memorize a formula for division, everybody just thinks of "rationalizing the denominator" (multiplying both numerator and denominator by the conjugate of the denominator):

$$(a+bi)/(c+di)=\big[(a+bi)/(c+di)\big]/\big[(c-di)/(c-di)\big]$$

$$= [(a+bi)(c-di)]/(c^2+d^2)$$

So, for example, $1/(1+i)=(1-i)/2=½-½i$.

With this as background, we can now define a *Gaussian integer* as a complex number $a+bi$ where a and b are *integers*, not just real numbers. Observe that any element n of $\mathbb{Z}$ is also an element of $\mathbb{Z}[i]$, simply because $n=n+0i$. We will often use the term "ordinary integer" to refer to the elements of $\mathbb{Z}$.

It follows readily from the foregoing that the set $\mathbb{Z}[i]$ is closed under the operations of addition, subtraction and multiplication but is not closed under the operation of division: the calculation above shows this because 1 and $1+i$ are both Gaussian integers but their quotient is not. In this respect, $\mathbb{Z}[i]$ is similar to $\mathbb{Z}$, the set of ordinary integers: the elements of $\mathbb{Z}$ can also be added, subtracted and multiplied, but not generally divided.

There is one respect in which the set $\mathbb{Z}[i]$ behaves somewhat differently than the set of ordinary integers: in the latter set, there is an order operation, but it is not possible to order the elements of $\mathbb{Z}[i]$. To see why, observe that, if we had such an order operation, then the square of any nonzero element would be positive, but of course $i^2=-1<0$.

However, although we can't speak of "positive" or "negative" Gaussian integers, we can define a *norm* on them that allows us some measure of comparison. If $\alpha=a+bi$ is a Gaussian integer, then the *norm* of α, denoted $\mathrm{N}(\alpha)$, is the integer a^2+b^2. (This is, of course, the same as $\alpha\bar{\alpha}$.) Observe that $\mathrm{N}(\alpha)$ is always a nonnegative integer and will be strictly positive if α is nonzero. Another useful property of the norm is given by the following theorem, the proof of which is very easy and is therefore omitted.

Theorem 7.1.1

If α and β are Gaussian integers, then $\mathrm{N}(\alpha\beta)=\mathrm{N}(\alpha)\mathrm{N}(\beta)$.

The Gaussian integers α that satisfy $\mathrm{N}(\alpha)=1$ are worth looking at in some detail. On the one hand, since the equation $a^2+b^2=1$ clearly has only four solutions in integers a and b (specifically $a=\pm1$, $b=0$ or $a=0$, $b=\pm1$), we see that the only four Gaussian integers that have norm 1 are 1, -1, i and $-i$. Observe that each of these four Gaussian integers has a multiplicative inverse in $\mathbb{Z}[i]$. Conversely, if α is any Gaussian integer with a multiplicative inverse β, then, taking norms on both sides of the equation $\alpha\beta=1$ and using the multiplicativity of the norm, we get $N(\alpha)N(\beta)=1$, from which it follows that $N(\alpha)=1$, and so α must be one of the four Gaussian integers specified above. We have thus proved:

Theorem 7.1.2

If α is a Gaussian integer, then the following are equivalent:

 (a) $N(\alpha)=1$
 (b) α has a multiplicative inverse in $\mathbb{Z}[i]$
 (c) $\alpha=1, -1, i$ or $-i$

If a Gaussian integer α satisfies any one (hence all) of the equivalent conditions above, it is called a *unit*. We say that Gaussian integers α and β are *associates* if $\alpha=\beta u$ for some unit u.

Exercises

 7.1 Express $(1-i)^4$ in the form $a+bi$.

 7.2 Is $(11-7i)/(1+3i)$ a Gaussian integer? Explain.

 7.3 Find, with proof, all ways to write $1+2i$ as the product $\alpha\beta$ of two Gaussian integers.

 7.4 Find all associates of $3-7i$.

 7.5 Prove that the relation "is an associate of" is an equivalence relation.

 7.6 Prove Theorem 7.1.1.

7.2 A Geometric Interlude

In this section, we interpret complex numbers and Gaussian integers geometrically, as points in a plane. We will use this interpretation to establish a fact, the significance of which will become apparent in Section 7.4, about the distance from any complex number to the closest Gaussian integer.

Any complex number $a+bi$ can be identified with the point (a, b) in the ordinary Cartesian plane. Indeed, this provides a way of actually defining a complex number that avoids reliance on the nebulous concept of an "imaginary unit", but we won't need this precise definition; we will, however, exploit the identification.

Under this identification, the Gaussian integers correspond to the points in the plane with two integer coordinates. Geometrically, these points form a lattice in the plane. They constitute the vertices of infinitely many "unit squares" that tile the plane, extending infinitely far from left to right and up and down.

Note that the distance from the point $z=a+bi$ to the origin is given by $|z| = \sqrt{(a^2+b^2)}$, and that for Gaussian integers z, $N(z) = |z|^2$.

Now let z be a complex number that is not a Gaussian integer. Then z lies in one (or two) of the unit squares that tile the plane—either in the interior, or on one of the sides. Pick a square containing z and call it S. Now divide S into four sub-squares of side length ½ by drawing the horizontal and vertical lines connecting the midpoints of the sides of S. The point z lies in at least one of these sub-squares, and each of these four sub-squares contains exactly one vertex of the original square S (i.e., contains exactly one Gaussian integer). The length of any diagonal of one of these sub-squares is, by the Pythagorean Theorem, equal to the square root of ¼+¼ = ½, or, putting it another way, $1/\sqrt{2}=\sqrt{2}/2<1$. It is clear, geometrically, that this is the largest possible distance from any point in a sub-square to the unique Gaussian integer (vertex) contained in that sub-square. Thus, we have shown the following geometric fact: *given any complex number z, there is a Gaussian integer α whose distance from z is less than 1; i.e., $|z-\alpha|<1$.* We will use this fact in Section 7.4, where we give a geometric proof of the Division Algorithm for the Gaussian integers.

Exercises

7.7   Find the Gaussian integer that is closest to the complex number $\frac{1}{2}+\frac{1}{4}i$.

7.8   Give an example to show that the Gaussian integer closest to a complex number z need not be unique.

7.3 Divisibility and Primes in the Gaussian Integers

Now that we have defined the Gaussian integers, we can begin to explore number theory in this system. Just as with the integers, we start with the basic idea of divisibility.

Definition 7.3.1

If α and β are Gaussian integers, then we say α divides β, denoted $\alpha|\beta$, if $\beta=\alpha\gamma$ for some Gaussian integer γ.

Just as with ordinary integers, the notion of divisibility satisfies certain basic properties. The ones listed below are the exact analogues of the ones specified in Theorem 1.2.1, except for parts (g) and (h), which follows immediately from the multiplicativity of the norm. The proofs of the other parts are also very simple and are therefore omitted.

Theorem 7.3.2

If α, β and γ are Gaussian integers, then:

(a) $\alpha \mid \alpha$
(b) $1 \mid \beta$
(c) if $\alpha \mid \beta$ and $\beta \mid \gamma$, then $\alpha \mid \gamma$
(d) if $\alpha \mid \beta$ and $\alpha \mid \gamma$, then $\alpha \mid \beta \pm \gamma$
(e) if $\alpha \mid 1$, then a is a unit
(f) *if* $\alpha \mid \beta$ and $\beta \mid \alpha$, then α and β are associates
(g) *if* $\alpha \mid \beta$ then $N(\alpha) \mid N(\beta)$
(h) if $\alpha \mid \beta$ and $N(\alpha)=N(\beta)$, then α and β are associates

Examples are easy to produce. Since $(3-i)(1+4i)=7+11i$, for example, it follows that $(3-i) \mid (7+11i)$. For a non-example, note that $(1+3i) / (1-3i)$ is not a Gaussian integer (check this!) and so $1-3i$ does not divide $1+3i$. Note also that since $1+3i$ and $1-3i$ obviously have the same norm, this last example also serves to establish that the converse of part (g) of the previous theorem is not true.

We can also adapt the definition of an ordinary prime integer to define prime elements in $\mathbb{Z}[i]$. We will use the word "irreducible" rather than "prime" to distinguish the notion from another kind of primality that will be discussed later. In the Gaussian integers these two ideas turn out to be equivalent, but that is not the case for certain other number systems, as we will see. We will use the Greek letter π to denote irreducible Gaussian integers, hoping that no confusion with the real number π will result.

Definition 7.3.3

A Gaussian integer π is called *irreducible* if it is not a unit, and if whenever $\pi=\alpha\beta$ for Gaussian integers α and β, then either α or β is a unit.

It has previously been observed that any ordinary integer is also a Gaussian integer. It should be kept in mind that the fact that an ordinary integer is prime does not mean that it is irreducible in the Gaussian integers. The integer 5 is certainly prime in $\mathbb{Z}$, but it is not irreducible in $\mathbb{Z}[i]$, as the factorization $5=(1+2i)(1-2i)$ shows.

We will ultimately state and prove a theorem that completely characterizes the irreducible Gaussian integers, but this will require the development of some more mathematical machinery. For the moment, however, we can at least record one easy theorem.

Theorem 7.3.4

If the norm of a Gaussian integer π is a prime integer, then π is irreducible.

Proof. Since $N(\pi)$ is prime, it certainly is not 1, so π is not a unit. Now suppose $\pi = \alpha\beta$. Taking norms on both sides gives us $N(\pi) = N(\alpha)N(\beta)$, and since $N(\pi)$ is a prime integer, this means either $N(\alpha)$ or $N(\beta)$ is equal to 1. Hence, either α or β is a unit.

The converse of this result is false. Consider, for example, the Gaussian integer 3. It has norm 9, which is not prime. Yet 3 is irreducible as a Gaussian integer. This follows from the next theorem.

Theorem 7.3.5

If p is an ordinary prime integer that is congruent to 3 mod 4, then p is an irreducible Gaussian integer.

Proof. Since p is an ordinary integer that is greater than 1, it is not a Gaussian unit. Suppose $p = \alpha\beta$, where neither α nor β is a unit. Taking norms gives $p^2 = N(\alpha)N(\beta)$. By uniqueness of prime factorization of ordinary integers, we must have $N(\alpha) = N(\beta) = p$. However, $N(\alpha)$ is, by definition of the norm, a sum of two squares, and a prime integer that is congruent to 3 mod 4 cannot be written as such a sum. (This follows from the simple observation that the square of any integer is congruent to 0 or 1 mod 4, and so the sum of two of them cannot be congruent to 3.) This contradiction yields the desired result.

Exercises

7.9   Prove that an associate of an irreducible is also an irreducible.

7.10 Factor both of the ordinary integers 26 and 44 into irreducibles in $\mathbb{Z}[i]$.

7.11 Prove that the ordinary integer n divides $a+bi$ if and only if n divides both a and b.

7.12 Prove that the conjugate of a Gaussian irreducible is also irreducible.

7.4 The Division Algorithm and the Greatest Common Divisor in $\mathbb{Z}[i]$.

The reader will recall that, when studying arithmetic in the ordinary integers, the Division Algorithm proved to be an indispensable tool. We begin this section, therefore, by stating and proving an analog of this result for the Gaussian integers, using the norm function as a way of bounding the remainder. The proof, thanks to the geometric result we established earlier, is refreshingly simple.

Theorem 7.4.1

Suppose that α and β are Gaussian integers, with $\beta \neq 0$. Then there exist Gaussian integers γ and ρ such that $\alpha = \beta\,\gamma + \rho$, and $0 \leq N(\rho) < N(\beta)$.

Proof. Consider α/β, which is not necessarily a Gaussian integer but which is certainly a complex number. By the geometric reasoning of Section 7.2, there is a Gaussian integer γ that satisfies $|\alpha/\beta - \gamma| < 1$, or (multiply through by $|\beta|$) the equivalent inequality

$|\alpha - \beta\gamma| < |\beta|$. Now define $\rho = \alpha - \beta\,\gamma$. Then it is obvious that $\alpha = \beta\,\gamma + \rho$, and moreover,

$N(\rho) = |\rho|^2 < |\beta|^2 = N(\beta)$. This completes the proof.

Note that this proof gives a method for actually computing the greatest common divisor of two Gaussian integers; the reader can try his or her hand at computing a gcd in Exercise 7.13. Note also that, in contrast to the situation for ordinary integers, the quotient and remainder here are not necessarily unique, because there may be more than one γ that is closest to α/β. For a simple example, note that if $\alpha = 1$ and $\beta = 1 - i$, then we have

$$1 = (1-i)0 + 1$$

$$= (1-i)1 + i$$

$$= (1-i)(1+i) + (-1)$$

$$= (1-i)(i) + (-i)$$

and in each equation above we have a "legitimate" quotient and remainder.

With the Division Algorithm in hand, we can discuss other aspects of divisibility in the Gaussian integers, just as we did for the ordinary integers. Our first task is to define a greatest common divisor of two Gaussian integers. (Note the use of the definite article "a" rather than "the"; we will see that in this broader context we have uniqueness only up to associates.) There are several ways to define a greatest common divisor. One way is to define it, as the name implies, as a common divisor that is greatest in norm among all common divisors. This definition has the advantage of making it clear that a greatest common divisor exists, but a disadvantage of not telling us everything we need to know. So, we will give a different definition, where the existence, though not obvious, can be proved.

Definition 7.4.2

Suppose that α *and* β are Gaussian integers, not both zero. Then a *greatest common divisor of* α *and* β is a Gaussian integer δ with the properties:

(a) $\delta \mid \alpha$ and $\delta \mid \beta$, and
(b) *if* $\delta' \mid \alpha$ and $\delta' \mid \beta$ then $\delta' \mid \delta$

Our first objective is to prove that a greatest common divisor (gcd) actually exists. There are several ways to do this; we will mimic the "ideal-theoretic" argument used to establish the greatest common divisor of two ordinary integers. We first restate the definition of an "ideal", this time in the context of the Gaussian integers.

Definition 7.4.3

A nonempty subset I of $\mathbb{Z}[i]$ is called an *ideal* if (a) whenever α *and* β are in I, so is $\alpha + \beta$, and (b) if α is in I and β is any Gaussian integer at all, then $\alpha\beta$ is in I.

So, just as with the set of ordinary integers, an ideal of $\mathbb{Z}[i]$ is a nonempty subset of $\mathbb{Z}[i]$ that is both closed under addition and "super-closed" under multiplication: not only is the product of two elements of I in I, but the product of an element of I and any other Gaussian integer is in I. Examples of ideals are easy to give: the two "trivial" ones are $\{ 0 \}$ and $\mathbb{Z}[i]$ itself. For a less trivial example, let α be any Gaussian integer and denote by $<\alpha>$ the set of all multiples $\alpha\beta$ of α. It is easy to verify that this is an ideal, called the *principal ideal generated by* α. Observe that this example includes the two previous ones as special cases, since clearly $\{ 0 \} = <0>$ and $\mathbb{Z}[i] = <1>$.

As was the case with the set of ordinary integers, the principal ideals in $\mathbb{Z}[i]$ are *all* the ideals.

Theorem 7.4.4

If I is any ideal of $\mathbb{Z}[i]$, then $I = <\alpha>$ for some Gaussian integer α.

Proof. If $I = \{ 0 \}$ then, as just observed, $I = <0>$, and we are done. So, suppose I contains at least one nonzero Gaussian integer. Then, by the Well-Ordering

Principle, I contains a nonzero element, call it β, of minimal (positive) norm. I claim $I = <\beta>$. Clearly, by "super-closure" of multiplication, $<\beta> \subseteq I$. For the reverse inclusion, let α be an arbitrary element of I. By the Division Algorithm (Theorem 7.6), we may write $\alpha = \beta\gamma + \rho$, where $0 \le N(\rho) < N(\beta)$. Because both α and β are in I and I is an ideal, it follows that $\rho = \alpha - \beta\gamma$ is also in I. If $\rho \ne 0$, then this would give us a nonzero element of I with a (positive) norm strictly smaller than the norm of β, a contradiction. Thus $\rho = 0$, and so $\alpha = \beta\gamma \in <\beta>$, finishing the proof.

This result has significance in abstract algebra: it says, intuitively, that an algebraic system in which we have an analog of the Division Algorithm is one in which every ideal is principal. In more advanced courses, we would phrase this as "every Euclidean domain is a Principal Ideal Domain".

We now use this result to prove the existence of a gcd of two Gaussian integers, and also to prove, at the same time, an additional fact about that gcd. We say that a Gaussian integer γ is a *linear combination of α and β* if we can write $\gamma = \alpha\sigma + \beta\tau$ for Gaussian integers σ and τ. Using this terminology, we now prove:

Theorem 7.4.5

If α and β are Gaussian integers, not both zero, then a gcd of α and β exists and is a linear combination of α and β.

Proof. Let $I = \{\alpha\sigma + \beta\tau : \sigma, \tau \in \mathbb{Z}[i]\}$ be the set of all linear combinations of α and β. It is obvious that I is nonempty (it contains both α and β), and it is easy to see (check this!) that I is an ideal. Therefore, I is principal, and therefore consists of multiples of a Gaussian integer, say δ. We will prove that δ is a gcd of α and β. Since it is obvious that δ is also a linear combination of α and β by the way it is defined, this will complete proof.

To show that δ is a gcd of α and β, we first observe that $\mathbb{Z}$ divides both α and β. This follows from the observation, made in the previous paragraph, that I contains both α and β, and every element of I is a multiple of δ by the way it is defined.

Finally, suppose δ' also divides both α and β. Then it is clear that δ' also divides any linear combination of α and β. But one such linear combination is δ itself. So $\delta' \mid \delta$, and this completes the proof.

It should be observed that, when working with the integers, the greatest common divisor of two integers was unique. That was because the gcd could be defined as a *positive* integer satisfying certain properties. There is, however, no notion of "positivity" for Gaussian integers, and so we must sacrifice complete uniqueness. We can, however, salvage a partial result, the proof of which we leave to the exercises: *if δ is a gcd of α and β, and δ' is a Gaussian integer, then δ' is also a gcd of α and β if and only if δ' and δ are associates.*

Just as with ordinary integers, we say that two Gaussian integers α and β are *relatively prime* if they have 1 as a gcd. Equivalent conditions are that

(a) α and β have no common divisor other than a unit, and (b) 1 is a linear combination of α and β. One can readily check that if π and α are Gaussian integers, with π irreducible and a non-divisor of α, then α and π are relatively prime.

The Euclidean Algorithm for finding the greatest common divisor of two ordinary integers can be adapted readily enough to find the greatest common divisor of two Gaussian integers, but doing these calculations seems like something of a chore, so we won't pursue this further. (But see Exercise 7.14 if you can't resist trying your hand at this.) We do note, however, that sometimes we can use norms to calculate the gcd without having to go through any algorithmic procedures. For example, consider the Gaussian integer $1+4i$ and its conjugate $1 - 4i$. They both have norm 17, and so any non-unit common divisor would have to have norm 17 as well. (Why?) However, a divisor of a Gaussian integer with the same norm must be an associate of that Gaussian integer, and it is easy to see that $1+4i$ and $1 - 4i$ are not associates of each other, so there can be no non-unit common divisor of these two Gaussian integers. Hence, these Gaussian integers are relatively prime.

We can now state and prove another analog of a useful ordinary integer divisibility theorem, Euclid's Lemma.

Theorem 7.4.6

(*Euclid's Lemma for Gaussian Integers*) If π, α and β are Gaussian integers with π irreducible, and $\pi \mid \alpha\beta$, then either $\pi \mid \alpha$ or $\pi \mid \beta$.

Proof. Suppose that it is not the case that $\pi \mid \alpha$. Then by the remark above, π and α are relatively prime, and we can write 1 as a linear combination of these two Gaussian integers:

$$1 = \pi \lambda + \alpha\mu$$

Multiplying both sides of this equation by β yields

$$\beta = \beta\pi \lambda + \alpha\beta\mu.$$

Each summand of the right hand side above is clearly divisible by π, and hence so is the left hand side. We have therefore shown that if it is not the case that $\pi \mid \alpha$, then it must be the case that $\pi \mid \beta$. This completes the proof.

The previous theorem extends easily by mathematical induction to the following result: if π *is irreducible and* $\pi \mid \alpha_1 \ldots \alpha_n$ then $\pi \mid \alpha_j$ for some j, $1 \leq j \leq n$.

We have now developed enough material to state and prove an analog, for the Gaussian integers, of the Fundamental Theorem of Arithmetic for ordinary integers. In what follows, when we speak of a "product of irreducibles", we implicitly allow for the product to have just one term—i.e., a single irreducible Gaussian integer is considered to be a product of irreducibles.

Theorem 7.4.7

Any nonzero, non-unit Gaussian integer can be expressed as a product of irreducibles, and this factorization is unique up to order and associates.

Proof. We first prove the existence of a factorization into irreducibles. Suppose, then, that there is a nonzero, non-unit, Gaussian integer that cannot be written as the product of irreducibles; by the Well-Ordering Principle, therefore, there is one such with minimal norm. Call this Gaussian integer α. Since α is not itself irreducible, it can be written as $\alpha = \beta \gamma$, where neither β nor γ is a unit. Since $N(\alpha) = N(\beta \gamma) = N(\beta)N(\gamma)$, it follows that both $N(\beta)$ and $N(\gamma)$ are strictly smaller than $N(\alpha)$, and hence, by the way α was chosen, it must be the case that both β and γ can be written as a product of irreducibles. But if this is the case, then clearly $\alpha = \beta \gamma$ can be as well, which is a contradiction. Therefore, every nonzero, non-unit Gaussian integer can be written as a product of irreducibles.

Next, we prove the uniqueness (up to order and associates) of this decomposition. Suppose that a nonzero, non-unit Gaussian integer α can be written in two ways as a product of irreducible Gaussian integers, say $\alpha = \beta_1 \ldots \beta_n = \gamma_1 \ldots \gamma_m$. Now, since it is obvious from this equality that β_1 divides $\gamma_1 \ldots \gamma_m$, it follows from the extended form of Euclid's Lemma that β_1 divides γ_j for some j, $1 \leq j \leq m$. Renumbering if necessary, we can assume $j=1$. Since β_1 is a non-unit and divides the irreducible element γ_1, it must be an associate of γ_1, say $\gamma_1 = \beta_1 v_1$ for some unit v_1. Dividing by β_1 then gives the equation $\beta_1 \ldots \beta_n = v_1 \gamma_2 \ldots \gamma_m$.

Repeat this process. We claim that $m=n$ and that each β_i is paired with one and only one γ_j. For, suppose that $n < m$. In that case, we will wind up with an equation with 1 on the left-hand side and a product of terms, one of them γ_{n+1}, on the right-hand side. But this would mean that γ_{n+1} divides 1, i.e., is a unit, which is a contradiction. A similar contradiction results if we assume that $m < n$. So we in fact have $n = m$, and a perfect pairing between irreducibles on the left-hand side and on the right-hand side, thus proving uniqueness (up to order and associates), as desired.

Exercises

7.13 Find a quotient and remainder when $3+2i$ is divided by $1-i$. Are they unique?

7.14 Find a gcd of $2+2i$ and $-3+5i$, and express this gcd as a linear combination of $2+2i$ and $-3+5i$.

7.15 Prove that two Gaussian integers are relatively prime if and only if there is no Gaussian irreducible that divides them both.

7.5 An Application: Sums of Two Squares

In this and the next two sections, we show how theorems about ordinary integers can be proved by invoking Gaussian integers. We first address the question of what integers can be written as the sum of two squares. (This question was alluded to in the exercises to Chapter 0.) We first answer this question for prime integers and then use that result to answer the question for all integers. It is easy to see that if an odd prime can be written as the sum of two integers, then that prime must be congruent to 1 mod 4. It takes a bit more work to prove the converse. We first start with an easy lemma.

Lemma 7.5.1

If a and b are positive integers that can each be written as the sum of two squares, then ab can as well.

Proof. Write $a=m^2+n^2$ and $b=r^2+s^2$. We could just write down an algebraic identify for ab, but that would be unmotivated; let us *discover* such a result by using Gaussian integers. If we let $\alpha = m+ni$ and $\beta = r+si$, then $ab=N(\alpha)\,N(\beta)=N(\alpha\beta)$, which, by definition, is the sum of two squares.

We now characterize those primes that can be written as the sum of two squares.

Theorem 7.5.2

The ordinary prime integer p can be written as the sum of two squares if and only if $p=2$ or p is congruent to 1 mod 4.

Proof. The "only if" direction is easy and amounts to recalling that the square of an integer is congruent to either 0 or 1 mod 4 depending on whether that integer is odd or even. We leave the details to the reader and instead prove the more challenging "if" direction. The number 2 is obviously a sum of two squares, so suppose that p is a prime that is congruent to 1 modulo 4. We know that -1 must be a quadratic residue mod p, so, for some x, $x^2 \equiv -1 \pmod{p}$. It follows that p divides x^2+1. Thinking of this in the Gaussian integers, this means that $p \mid (x+i)$ $(x-i)$. Now, if p were irreducible in $\mathbb{Z}[i]$, this would imply $p \mid (x+i)$ or $p \mid (x-i)$. But since p is an ordinary integer, Exercise 7.11 clearly makes this impossible. Thus, p (viewed as a Gaussian integer) is not irreducible, which means we can write $p=\alpha\beta$, where neither α nor β is a unit. It follows, upon taking norms, that $N(\alpha)=p$. But if we write $\alpha =a+bi$, this means that $p=a^2+b^2$; i.e., that p is a sum of two squares, as was to be proved.

Using this, we can characterize all positive integers that are the sum of two squares. This argument does not use the Gaussian integers, but it does use a result that we established when we discussed quadratic reciprocity, namely that -1 is not a quadratic residue mod p for any prime p that is congruent to 3 mod 4.

Theorem 7.5.3

An integer $n > 1$ can be written as the sum of two squares if and only if every prime factor of n that is congruent to 3 mod 4 appears with even multiplicity in the prime factorization of n.

Proof. If this condition is satisfied, then n can be written as $s^2 t$ where s and t are positive integers and t is the product of distinct primes, each one either 2 or congruent to 1 mod 4. It follows immediately from Theorem 6.5.2 that t can be written as the sum of two squares, say $t = a^2 + b^2$. But then $n = (sa)^2 + (sb)^2$ is also a sum of two squares, as desired.

For the converse, suppose n is the sum of two squares, say $n = a^2 + b^2$. Let p be a prime dividing n that is congruent to 3 mod 4. We will show that p appears to an even power in the prime factorization of n. To do this, first note that since $p = a^2 + b^2$, we must have $a^2 \equiv -b^2 \pmod{p}$. If it were the case that p did not divide b, then b would be relatively prime to p and would have a multiplicative inverse mod p; multiplying both sides of the congruence $a^2 \equiv -b^2 \pmod{p}$ by that multiplicative inverse, we would see that -1 was a quadratic residue mod p, which we know it is not. So p divides b, and from this it is immediate that p divides a as well. It follows that p^2 divides both a^2 and b^2, and hence p^2 divides n.

If p^2 is the largest power of p dividing n, then we are done; if not, then p divides n/p^2. However, n/p^2 is also a sum of two square integers: $(a/p)^2 + (b/p)^2$. By what we have just shown, p^2 divides n/p^2, or p^4 divides n. If this is the largest power of p that divides n, we are again done; otherwise, repeat the process once more. The point is that we have to stop at some point, and we must stop at an even power of p, since every time that p divides n/p^k, so does p^2. This completes the proof.

Now take a prime, like 5, which is a sum of two squares: $5 = 2^2 + 1^2$. It is clear that, except for the order of the terms 2^2 and 1^2, this is the *only* way 5 can be written as the sum of two squares. The same is true of other primes like $13 = 2^2 + 3^3$ or $17 = 1^2 + 4^2$. The next result says that this is not a coincidence.

Theorem 7.5.4

If an ordinary integer prime p can be written as the sum of two squares of integers, then, except for the order of the terms, it can be done so in only one way.

Proof. Suppose the ordinary prime $p = a^2 + b^2 = c^2 + d^2$. Factoring this equation in the Gaussian integers gives $(a+bi)(a-bi) = (c+di)(c-di)$. Each of the four Gaussian integers that appear in this equation are irreducible, because each of them has norm p, a prime. By uniqueness of factorization into irreducibles, $a+bi$ must be an associate of either $c+di$ or $c-di$. Assume, for sake of argument, that $a+bi$ is an associate of $c+di$. Then $a+bi$ is either $c+di$, $-(c+di)$, $i(c+di)$ or $-i(c+di)$. In the first case, $a=c$ and $b=d$, in the second case $a=c$ and $b=-d$, in the third case $a=-d$ and $b=c$, and in the final case $a=d$ and $b=-c$. In all cases, the set of integers $\{a^2, b^2\}$ is just the set $\{c^2, d^2\}$. If $a+bi$ is an associate of $c-di$, nothing more needs to be done, because we have just changed d by a sign, which does not affect its square. This proves the result.

Note that if p is *not* a prime, this result is not necessarily true. For example, $65 = 49+16 = 1+64$.

We end this section by addressing some questions that Theorem 7.5.3 naturally raises: What about sums of three squares? What about sums of four squares? The answers to both of these questions are known, though the proof for sums of three squares is fairly difficult and not especially fun, so we omit the proof. The answer, though, is that a positive integer n can be written as the sum of three nonnegative squares if and only if n is not of the form $4^k(8m+7)$ for nonnegative integers k and m.

The answer to the second question is more interesting: it turns out that *any* positive integer can be written as the sum of four nonnegative squares. (You may have guessed this if you did Exercise 0.3.) We will prove this result later in the chapter by exploiting another arithmetic system that generalizes the integers (and in fact generalizes the Gaussian integers).

Exercises

7.16 Determine whether each of the integers 688, 1000 and 1240 can be written as the sum of two squares.

7.17 The last paragraph of the proof of Theorem 6.5.3 is a little informal. Make it precise by showing that, in the notation of the theorem, p^{2k+1} cannot be the largest power of p that divides n.

7.18 Prove that, among any four positive consecutive integers, at least one cannot be written as the sum of two squares.

7.19 Prove that if p is a prime that is congruent to 3 mod 4, then p^2 is not a sum of two *positive* squares.

7.6 Another Application: Diophantine Equations

As a second application of Gaussian integers to the ordinary integers, we show how Gaussian integers can also be used to help determine the solutions to Diophantine equations. These are polynomial equations with integer coefficients, for which integer solutions are sought. We illustrate this idea with the Diophantine equation $y^2 = x^3 - 1$. One solution, which we can find by inspection, is $x=1$ and $y=0$. It turns out that this is the only solution, and Gaussian integers can be used to prove this.

Theorem 7.6.1

The equation $y^2 = x^3 - 1$ has $x=1$ and $y=0$ as its only solution.

Proof. Suppose that x and y are integer solutions to the equation. Then, working in $\mathbb{Z}[i]$, we have $x^3 = y^2 + 1 = (y+i)(y-i)$. We first claim that the two terms on the right hand side are relatively prime. Assuming this for the moment, we will finish the proof.

It is an easy consequence of unique factorization in $\mathbb{Z}$ that if the product mn of two relatively prime positive integers m and n is a cube, then each term in the product is a cube. To see this, observe that every prime in the factorization of mn must have exponent divisible by 3; since the prime factorizations of each term in the product cannot have a prime in common, it follows that each prime in the factorization of m and n also has exponent divisible by 3. We would like to adopt this reasoning for $\mathbb{Z}[i]$ and assert that if the product of two relatively prime Gaussian integers is a cube, then so is each term in the product, but a subtle point intervenes: all we can say here is that each term in the product is an associate of a cube. However, an associate of a cube is itself a cube, since all four units in $\mathbb{Z}[i]$ are themselves cubes in $\mathbb{Z}[i]$. Since the product of two cubes is a cube, it follows that this result does indeed carry over to $\mathbb{Z}[i]$.

So, based on the equation in the first paragraph above, it must be the case that $y+i$ is a cube, say, $y+i = (m+ni)^3$. If we expand the right-hand side and equate the imaginary terms, we get the equation $1 = n(3m^2 - n^2)$. This tells us that $n=1$ or $n=-1$. If $n=1$, then $1 = 3m^2 - 1$, which is impossible for any integer m. If $n=-1$, then we get $1 = -(3m^2 - 1)$ or $m=0$. Thus $m=0, n=-1$. Since $y+i = (m+ni)^3$, it follows that $y=0$, from which we immediately conclude that $x=1$.

We are thus done, given our assumption that the terms $y+i$ and $y-i$ are relatively prime. We show this as follows. Suppose to the contrary that a Gaussian integer π divides both of these terms. Then π divides their difference, which is $2i$. Since i is a unit, this means that π divides $2 = (-i)(1+i)^2$. Hence (by unique factorization) if π is not a unit, it is an associate of either $1+i$ or $(1+i)^2$. In either case, $1+i$ divides π. This, in turn, means that $1+i$ divides x^3. Taking norms, we see that x^6, and hence x, must be even. Now return to the equation $y^2 = x^3 - 1$; reading it mod 4 tells us that y^2 must be congruent to -1 (or 3) mod 4. But this is a contradiction, since no square is congruent to 3 mod 4. This contradiction yields the desired result and finishes the proof.

The result established above can be generalized: the equation $y^2 = x^d - 1$ has $(1,0)$ as its only solution for any $d \geq 3$. This was proved by Lebesgue, also using the Gaussian integers.

Exercises

7.20 Suppose we switch the roles of y^2 and x^3 in the Diophantine equation studied above and consider the equation $x^3 = y^2 - 1$. Does the uniqueness result established above still hold?

7.21 Look up *Catalan's Conjecture* and write a brief essay about it.

7.7 A Third Application: Pythagorean Triples

In Section 2.5, we characterized Pythagorean Triples. We did so then using an elementary but not terribly exciting elementary argument using nothing more than basic divisibility properties. We now give a more interesting argument using Gaussian Integers. Because the reader should, by now, have some familiarity with the techniques that will be used, we leave parts of the verification of this result as exercises.

Recall that it suffices to consider primitive Pythagorean triples—i.e., positive integers a, b and c, not having any positive common divisor other than 1, satisfying the identity $a^2 + b^2 = c^2$. Recall also that we can assume without loss of generality that a is odd and b is even. The relevant theorem, which we reproduce here for convenience, states:

Theorem 7.7.1

Let a, b and c be three relatively prime positive integers, with a odd, b even, and $a^2 + b^2 = c^2$. Then there exist positive, relatively prime, integers m and n of opposite parity such that $a = m^2 - n^2$, $b = 2mn$ and $c = m^2 + n^2$.

Proof. The equation $a^2 + b^2 = c^2$ factors, in the Gaussian integers, as $(a+bi)(a-bi) = c^2$. The first thing to observe is that, as in the proof of the preceding result, $(a+bi)$ and $(a-bi)$ are relatively prime. The proof is not too different than the proof given in the previous result, and we leave it as an exercise.

Now that we know that the product of two relatively prime Gaussian integers is equal to a square, it is tempting to assert that each term is a square. We used similar reasoning, with cubes replacing squares, in the preceding proof. But as noted in that proof, there's a subtle point here: since irreducible factorization is unique *only up to associates*, all we can really conclude is that each term is an associate of a square. This didn't create an issue in the previous result, because every unit in the Gaussian integer is a cube. However, every unit in the Gaussian integers is *not* a square (specifically, i and $-i$ are not), so things are not quite as simple now. We can therefore, a priori, only assert that $a + bi = (m+ni)^2$ or $a + bi = i(m+ni)^2$. (Why do we not need to consider the case $a + bi = (-i)(m+ni)^2$?)

If the second case holds, then, expanding the square and equating real parts, we get $a = -2mn$, which contradicts our assumption that a is odd. So in fact this case cannot hold after all.

We now know, then, that $a + bi = (m + ni)^2$. Squaring the right-hand side and equating real and imaginary parts gives us *that* $a = m^2 - n^2$ and $b = 2mn$, as desired. Observe that m and n can chosen to be positive; they are obviously either both positive or both negative (why?), and in the latter case, we may replace each one by its negative. Also, note that m and n must be relatively prime, because a and b are. They must also be of opposite parity, as otherwise a and b would both be even, a contradiction. Finally, computing $c^2 = a^2 + b^2$, then gives us $c = m^2 + n^2$, completing the proof.

Exercises

7.22 Prove that, in the proof of Theorem 7.7.1, the Gaussian integers $(a + bi)$ and $(a - bi)$ are relatively prime.

7.23 Explicitly answer the question posed in the proof above: why do we not need to consider the case $a + bi = (-i)(m + ni)^2$?

7.8 Irreducible Gaussian Integers

In this section, we classify all irreducible elements in $\mathbb{Z}[i]$. We already know some of them: Gaussian integers with prime norm, and ordinary prime integers that are congruent to 3 modulo 4. Our next theorem provides a complete list.

Theorem 7.8.1

A Gaussian integer is irreducible if and only if it is an associate of one of the following:

(a) $1 + i$
(b) a Gaussian integer π, where $N(\pi)$ is an ordinary prime congruent to 1 mod 4
(c) an ordinary prime p that is congruent to 3 mod 4

Proof. We already know that every Gaussian integer described above is irreducible, so it suffices to prove the converse. Let π denote an irreducible Gaussian integer. We first show that there is an ordinary prime integer p that is a multiple of π. This is easy: observe that π divides $N(\pi) = \pi\bar{\pi}$, and that $N(\pi)$, being a positive integer greater than 1, is a product of ordinary integer primes. Because π divides this product, it must, by Euclid's Lemma, divide one of the ordinary integer primes making up that product.

It follows from the previous paragraph that $N(\pi)$ divides $N(p)=p^2$. Since π is not a unit and therefore cannot have norm 1, it follows that $N(\pi)$ must be either p or p^2. We consider both cases in turn. First suppose that $N(\pi)=p$. If $p=2$, this clearly forces π to be one of $1+i$, $1-i$, $-1+i$ or $-1-i$. All of these numbers, however, are associates of $1+i$, so in this case we are done. If p is odd, then it must be congruent to 1 mod 4, and therefore, π falls into case (b) above, and we are also done.

Finally, suppose that $N(\pi)=p^2$. Since π divides p and $N(\pi)=N(p)$, it follows from part (h) of Theorem 7.3 that π is an associate of p. Note also that because 2 (= $(1+i)(1-i)$) is not irreducible, we must have $p \neq 2$, in which case p must be congruent to 3 mod 4: The only other possibility would be for p to be congruent to 1 mod 4, in which case p would be the sum of two squares, say $p=a^2+b^2$, but then $p=(a+bi)(a-bi)$ would not be irreducible. So, in this remaining case, π falls into case (c) above, and we are done.

Now that we know what irreducibles in the Gaussian integers look like, we can get some practice in actually writing a Gaussian integer as a product of irreducibles. Let us start, for example, with the Gaussian integer 6. We can certainly write this as $2 \cdot 3$, which would be a prime factorization in the ordinary integers, but 2 is not irreducible in $\mathbb{Z}[i]$; it factors as $(1+i)(1-i)$, both of which are irreducible. The ordinary integer 3, viewed as a Gaussian integer, *is* irreducible (type (c) above). So the "Gaussian prime factorization" of 6 is $3(1+i)(1-i)$.

Now let's work with a non-real Gaussian integer, say $\alpha=1+7i$. By now we should expect that the norm of α should likely play a role in finding its irreducible factors. The norm of α is, in fact, 50, which factors (in the ordinary integers) as 2 times 5^2. Therefore, if α has irreducible factors, we expect that these factors will have norms 2 and 5. We already know a Gaussian integer with norm 2, namely $1+i$; it is easy to see that any other Gaussian integer with norm 2 must be an associate of this one. So, we expect that $1+i$ is a divisor of $1+7i$, and we verify this by simple division: $\alpha=(1+i)(4+3i)$. (This is not a coincidence. In the exercises, you will show that any Gaussian integer with even norm is divisible by $1+i$.)

We are not done, however, because the last factor above, $4+3i$, is not irreducible; it does not fit any of the three categories of the previous theorem. Since its norm is 25, we would expect it to factor into two Gaussian integers, each with norm 5. One possible candidate is $1-2i$; however, when we divide $4+3i$ by $1-2i$, we do not get a Gaussian integer (check this!). So $1-2i$ cannot be a factor of $4+3i$; neither can, therefore, any associate of $1-2i$. But the *conjugate* of $1-2i$, $1+2i$, is not an associate of $1-2i$, and also has norm 5. Let's try this. Here, we get lucky: a calculation shows that $(4+3i)/(1+2i)=2-i$. So, we have computed that $\alpha=(1+i)(4+3i)=(1+i)(1+2i)(2-i)$. Each of these three factors is irreducible, because they all have prime norm. So now we are done: we have found a factorization of $1+7i$ into irreducible factors.

Exercises

7.24 Explain why the factorization $10 = 2 \cdot 5 = (1+3i)(1-3i)$ does not contradict unique factorization into irreducible elements in $\mathbb{Z}[i]$.

7.25 Express $4+12i$ as a product of irreducible elements in $\mathbb{Z}[i]$. Do the same for the Gaussian integer $-101i$.

7.26 Factor the ordinary integer 65 as a product of irreducible elements in $\mathbb{Z}[i]$. Recall that in Section 6.5 we pointed out that this integer can be expressed as a sum of squares in two different ways. Explain how, by rearranging the terms of the irreducible factorization that you just found, this result could be discovered.

7.27 Prove that a Gaussian integer α is divisible by $1+i$ if and only if $N(\alpha)$ is even.

7.28 In the proof of Theorem 6.8.1, we first showed that that there is an ordinary prime integer p that is a multiple of π. Is p unique? Prove your answer.

7.29 Suppose that (x, y, z) is a primitive Pythagorean triple. Prove that z is never divisible by a prime that is congruent to 3 mod 4.

7.9 Other Quadratic Extensions

Up to this point, this chapter has been concerned with the Gaussian integers—an arithmetic system (or "ring", to use the technical algebraic term mentioned in Appendix C) obtained from the integers by "adjoining" the single element i. Since we want to add and multiply in our new system, once we add i we also have to add all elements of the form $a+bi$, thus giving us the Gaussian integers. The number i is a quadratic number; this notion can be made more precise in more advanced courses, but for our purposes it is enough to note that it is the square root of an integer, namely -1. There is no particular reason why we can't choose some other quadratic number, real or complex, and see what happens when we adjoin that to the integers. We will do that in this section, which is intended as a survey rather than an in-depth discussion. The focus will be on examples, rather than proofs.

To start, consider $\sqrt{2}$. If we want to have an arithmetic system containing all integers and this number, we must have in it all real numbers of the form $a+b\sqrt{2}$, where a and b are arbitrary integers, and the set of all such numbers is indeed closed under addition and multiplication (though not division). Let us denote this set $\mathbb{Z}[\sqrt{2}]$. We can define divisibility and irreducibility exactly as we did for Gaussian integers.

By analogy with the Gaussian integers, we can define a norm on this set: $N(a+b\sqrt{2}) = (a+b\sqrt{2})(a-b\sqrt{2}) = a^2 - 2b^2$. A short calculation shows that this norm, like the one defined on the set of Gaussian integers, is multiplicative, although here it should be noted that the norm may take on negative values. This requires modification of a theorem about the Gaussian integers: now, an element α in $\mathbb{Z}\left[\sqrt{2}\right]$ is a unit if and only if $N(\alpha)=\pm 1$. We leave the details of proving this to the exercises.

It can be shown that the equation $a^2-2b^2=\pm 1$ has infinitely many integer solutions. So, $\mathbb{Z}\left[\sqrt{2}\right]$ has infinitely many units, unlike the Gaussian integers, which has only 4. However, in another respect, $\mathbb{Z}\left[\sqrt{2}\right]$ is similar to the Gaussian integers: there is an analog of the Division Algorithm in $\mathbb{Z}\left[\sqrt{2}\right]$, and it follows from this that there is unique factorization into irreducible elements here as well. We will not prove these facts here.

Now let's vary things and consider $\mathbb{Z}\left[\sqrt{-2}\right]$, or the set of all complex numbers of the form $a+b\sqrt{2}i$. This set is also closed under addition and multiplication, and hence we can do basic arithmetic in this domain. In particular, we can define divisibility and irreducibility just as with the Gaussian integers. We can define a norm on elements of this set by exact analogy with the Gaussian integers: $N(a+b\sqrt{2}i)=(a+b\sqrt{2}i)(a-b\sqrt{2}i)=a^2+2b^2$. So here, the norm takes on only positive values, is multiplicative and satisfies $N(a+b\sqrt{2}i)=1$ if and only if $a+b\sqrt{2}i$ is a unit. The equation $a^2+2b^2=1$, however, obviously has as its only solutions $a=\pm 1$, $b=0$, so this ring has only two units: ± 1. It can be shown, though we won't do so here, that an analog of the Division Algorithm holds for this ring, and, with it, unique factorization into irreducibles.

We next consider $\mathbb{Z}\left[\sqrt{-5}\right]$, or the set of all numbers of the form $a+b\sqrt{5}i$. We define the norm of this generic element to be a^2+5b^2, and, as in the previous case, this is equal to 1 only when $a=\pm 1$, $b=0$, so this ring has only two units: ± 1. But $\mathbb{Z}\left[\sqrt{-5}\right]$ differs from $\mathbb{Z}\left[\sqrt{-2}\right]$ in one very important respect: this time, there is no Division Algorithm, and unique factorization into irreducibles fails. In fact, we will show that $6=3\times 2=(1+\sqrt{5}i)(1-\sqrt{5}i)$ gives two distinct irreducible factorizations of 6.

To see this, observe that no one of the four elements that appear as factors of 6 is an associate of any other one. We can also show that each of these four elements is irreducible. Suppose, for example, that $2=\alpha\beta$, where α and β are non-unit elements of $\mathbb{Z}\left[\sqrt{-5}\right]$. Taking norms gives $4=N(\alpha)N(\beta)$. This in turn implies, by unique factorization of ordinary integers, that $N(\alpha)=N(\beta)=2$. But this is impossible, because 2 cannot be written as $=a^2+5b^2$. For precisely the same reasons, both $(1+\sqrt{5}i)$ and $(1-\sqrt{5}i)$ are also irreducible. Thus, we have two distinct (even up to associates) factorizations of 6 into irreducible elements.

It is worthwhile to consider the equation $3\cdot 2=(1+\sqrt{5}i)(1-\sqrt{5}i)$ from the standpoint of Euclid's Lemma. Notice that the irreducible element 2 divides the product on the right-hand side but does not divide either of the two terms making up this product—i.e., Euclid's Lemma for the integers fails for some quadratic extensions of the integers. It is for this reason that many books distinguish, in these extensions, the concepts of *prime* and *irreducible* element. Irreducible elements have already been defined; a prime element p is a non-zero, non-unit element with the property that whenever p divides a product ab, it must divide either a or b. So we have now seen an example of an irreducible element that is not prime; the converse, however, cannot occur; the proof of this is an exercise.

In our discussion so far, we have looked at quadratic extensions of the integers. It is worth noting that, although we will not study the subject here, we can consider other kinds of extensions of the integers as well. We limited our discussion to quadratic extensions simply because these are the simplest examples.

The fact that unique factorization may fail in extensions of the integers has considerable historical and mathematical significance. Historically, for example, the incorrect assumption of unique factorization in such extensions led to a famous faulty proof of Fermat's Last Theorem. Recall that this famous conjecture stated that the equation $x^n + y^n = z^n$ has no solutions in positive integers if $n > 2$. This can be proved directly if $n = 4$, and, it is therefore easy to see that if it were known to also be true for all odd primes n, then it would be true for all $n > 2$. So, assume n is an odd prime.

In 1847, the mathematician Gabriel Lame announced that he had found a proof. His proof involved assuming a solution, and then factoring the left-hand side of the above equation, not in the ring of integers, but in the larger (cyclotomic) ring $\mathbb{Z}[\zeta]$ obtained by adjoining a primitive nth root of unity ζ to the ring of integers. In $\mathbb{Z}[\zeta]$, the left-hand side above factors as the product of all terms $x + \zeta^i y$ as i ranges from 0 to $n - 1$. Lame argued that it could be assumed that all the terms in this product were relatively prime, and, using the principle that the product of relatively prime terms is a prime power if and only if each term is, derived a contradiction.

The problem, of course, is that not every such extension of the integers satisfies the unique factorization property. This was pointed out by Kummer.

It should not surprise the reader to hear that the examples described above constitute the very tiniest tip of a rather large iceberg. The (possible) failure of unique factorization into irreducibles can be frustrating, but it is also an opportunity, and the study of this phenomenon has enhanced the areas of mathematics known as commutative algebra and algebraic number theory. In the next section, we will elaborate on this comment, in a purely expository way without filling in the details or offering proofs. Some prior familiarity with the notion of a polynomial is assumed for this next section.

Exercises

7.30 Prove that an element α in $\mathbb{Z}\left[\sqrt{2}\right]$ is a unit if and only if $N(\alpha) = \pm 1$.

7.31 Give an example to show unique factorization into irreducibles fails in $\mathbb{Z}(\sqrt{-14})$.

7.10 Algebraic Numbers and Integers

The numbers that we have been considering, such as $1 - \sqrt{5}i$, are examples of algebraic numbers. An *algebraic number* is a complex number α that is the root of a nonconstant monic polynomial (i.e., a polynomial whose highest term has coefficient 1) with rational coefficients. By clearing denominators, we see that this definition is equivalent to the requirement that α is the root of a nonconstant polynomial (not necessarily monic) with integer coefficients. If α is the root of a nonconstant *monic* polynomial with integer coefficients, then we say that α is an *algebraic integer*. A complex number that is not an algebraic number is called *transcendental*.

Some examples: any rational number r is an algebraic number because it is a root of the polynomial $x - r$. The number $\sqrt{2}$ is irrational but is algebraic, because it is the root of the polynomial $x^2 - 2$; this observation also shows that it is an algebraic integer. The numbers π and e were shown to be transcendental in 1882 (by Lindemann) and 1873 (by Hermite), respectively.

Although it is not obvious, it can be shown (using ideas from abstract algebra) that if α and β are algebraic numbers, then so are $\alpha + \beta$, $\alpha - \beta$, $\alpha\beta$ and α/β (assuming that $\beta \neq 0$). To use algebraic terminology, the set of algebraic numbers is a subfield of the set of complex numbers. The set of algebraic integers, however, is not, as we see in the exercises.

Every algebraic number α has a *degree*, which can be defined as follows: since α satisfies a nonconstant monic polynomial with rational coefficients, it satisfies one of least degree. One can show that this polynomial is irreducible (i.e., that it cannot be nontrivially factored into two polynomials with rational coefficients). This polynomial is called the minimal polynomial of α; the *degree of* α is then defined to be the degree of its minimal polynomial. So, the degree of $\sqrt{2}$ is 2. We now look at a class of algebraic numbers of degree 2.

If d is a squarefree integer (i.e., an integer not divisible by a square other than 1), we can then define the set $\mathbb{Q}(\sqrt{d})$ to be the set of numbers of the form $a + b\sqrt{d}$, where a and b are rational numbers. Note that we allow d to be negative, so these numbers may be complex. It follows from the previous paragraphs that every number of this form is an algebraic number. In fact, simple calculation shows that every element of this set is an algebraic number of degree 1 or 2. It is not hard to show that the set $\mathbb{Q}(\sqrt{d})$ is also closed under the four basic operations of addition, subtraction, multiplication and division (by nonzero elements), so it is also a subfield of the set of complex numbers.

Suppose we now ask: what are the algebraic integers in the set $\mathbb{Q}(\sqrt{d})$? It is tempting to guess that the set of algebraic integers is the set $\mathbb{Z}(\sqrt{d})$, but in fact that turns out to not always be the case. The precise answer is given by this theorem, which we state without proof:

Theorem 7.10.1

Let d be a squarefree integer. If d is congruent to 2 or 3 mod 4, then the algebraic integers in $\mathbb{Q}$ ($\sqrt{d}$) are of the form $a + b\sqrt{d}$, where a and b are integers. If, on the other hand, d is congruent to 1 mod 4, then the algebraic integers in $\mathbb{Q}$ ($\sqrt{d}$) are of the form $a + (b/2)(1 + \sqrt{d})$ for integers a and b.

So, for example, if $d = -5$, then the set of algebraic integers $\mathbb{Z}$ ($\sqrt{d}$) fails to have unique factorization into irreducibles. In an attempt to recover something of unique factorization, mathematicians like Dedekind and Kummer invented the notion of an *ideal*: one can define a product of ideals and show that even though there is not unique factorization into irreducibles, there is unique factorization into prime ideals. The area of mathematics known as *algebraic number theory* elaborates on these ideas, but that is a subject that is beyond the scope of this text. A good introductory reference to this subject is [Jar].

Exercises

7.32 Show, from the definition, that $\sqrt{2} + \sqrt{3}$ is an algebraic integer, as is $1 + \frac{1}{2}(1 + \sqrt{-3})$.

7.33 Find two "essentially different" factorizations of 4 into irreducible elements in the set $\mathbb{Z}$ ($\sqrt{-3}$).

7.34 The result of the preceding problem may suggest that the set of algebraic integers in $\mathbb{Q}(\sqrt{-3})$ does not have unique factorization into irreducible elements, but in fact it does, though we won't prove this. The reason is that, by virtue of Theorem 7.10.1, $\mathbb{Z}$ ($\sqrt{-3}$) is the wrong set to look at. Find all units in the full set of algebraic integers of $\mathbb{Q}$ ($\sqrt{-3}$), and show that in the factorizations you gave in response to the previous question, the irreducible terms in one factorization are associates of the irreducible terms in the other.

7.11 The Quaternions

The set of Gaussian integers that we have studied for most of this chapter consisted of complex numbers, which are generalizations of the real numbers. We now introduce a set of numbers that generalize the complex numbers. The complex numbers were obtained from the real numbers by the adjunction of an "imaginary unit" i; we now add two additional "imaginary units" that we will call j and k. These can be thought of as formal symbols that satisfy $j^2 = k^2 = -1$. Why add two new symbols instead of just one? It turns out that this is necessary to obtain a satisfactory arithmetic system, but answering this question in more specific detail would require us to wade in deeper mathematical waters than we are prepared to here. We call the resulting set of numbers the *quaternions*.

Thus, the set of quaternions consists of all expressions of the form $a+bi+cj+dk$, where a, b, c and d are real numbers and $i^2=j^2=k^2=-1$. Addition of two quaternions, like addition of two complex numbers, is defined "component-wise": just add corresponding real coefficients. To define multiplication, we first explain how i, j and k multiply together: we define $ij=k$, $jk=i$, $ki=j$, $ji=-k$, $kj=-i$, $ik=-j$. With these definitions, we can then define multiplication of two quaternions by using these relations and the associative and distributive law, along with the fact that the real numbers commute with everything.

The set of quaternions is denoted $\mathbb{H}$, in honor of the Irish mathematician William Rowan Hamilton, who discovered them in 1843. The story behind this is interesting: Hamilton was talking a walk in Dublin and was passing by the Brougham Bridge when he suddenly realized how these numbers should be defined. He was so delighted with his discovery that he scratched the defining relations onto the bridge.

Although we won't go through the computational details here, it turns out that in the set $\mathbb{H}$ we can add, subtract, multiply and divide and that these operations satisfy all but one of the familiar properties of these operations. The one exception is that multiplication is no longer commutative. This is in fact apparent from the defining relations specified above, since, for example, $ij \neq ji$; the left-hand side is the negative of the right-hand side. So $\mathbb{H}$ is not a field (see Appendix C), but it is what is called a *skew-field* or *division ring*.

As in our previous work, we want to define the "integers" in $\mathbb{H}$. Our first guess would be to say that a quaternion $a+bi+cj+dk$ is an integer if and only if each of the coefficients a, b, c and d are, but, as we saw earlier in this chapter, the "obvious" definition is not always the "right" one. It turns out in this case that a better definition is as follows: an integral quaternion $a+bi+cj+dk$ is one where each of the coefficients a, b, c and d are integers or each of these coefficients are "half integers": i.e., each one is of the form $n+\frac{1}{2}$ for some integer n. Another way to say this is to say that each of $2a$, $2b$, $2c$ and $2d$ are integers and these all have the same parity. These integral quaternions are called Hurwitz integers; we will denote the set of them by $\mathcal{O}$.

We can offer a brief explanation as to why we need to consider half-integers as well as integers. Recall that, in the case of Gaussian integers, we proved the existence of a Division Algorithm by proving that, given any complex number z, there was a Gaussian integer whose distance from z was strictly less than 1. This, in turn, was proved by a geometric argument, using the fact that Gaussian integers formed the vertices of squares. Now we must deal with four dimensions instead of two, but there is still a geometric observation to be made: the quaternions with integer coefficients form the vertices of "hypercubes". The problem, however, is that it is no longer the case that one of these vertices has length less than 1 from an arbitrary quaternion. For example, the quaternion $\frac{1}{2}+\frac{1}{2}i+\frac{1}{2}j+\frac{1}{2}k$ has distance exactly 1, not less than 1, from the nearest integer-coefficient quaternion vertex. This observation allows us to conclude that a good Division Algorithm does not hold in the set of integer-coefficient quaternions.

If $q=a+bi+cj+dk$ is a quaternion, then we define the *conjugate* of q, denoted $\bar{q}$, to be the quaternion $a-bi-cj-dk$. A tedious calculation (which we omit) shows that for quaternions q and w, we have $\overline{qw}=\bar{w}\,\bar{q}$. (Note that we have switched the order of q and w—remember that quaternion multiplication is not commutative!) We define the *norm* of q, N(q), to be $q\,\bar{q}$, which, a calculation shows, is just $a^2+b^2+c^2+d^2$.

It follows easily from the definition above that the norm is multiplicative, just as it was for the Gaussian integers. We leave the verification of this as an exercise. It is also valuable to note that if q is in fact a Hurwitz integer, then N(q) is an ordinary integer. This is obvious in the case where a, b, c and d are themselves ordinary integers; if they are each half-integers, then the result follows either from a brute force calculation or from the cleverer observation, using the penultimate sentence of the fourth paragraph of this section, that $4(a^2+b^2+c^2+d^2)$ must be an integer that is congruent to 0 mod 4. We also leave the details of this argument to the exercises.

It would be nice if we could define divisibility in the set $\mathcal{O}$ just as it was defined for Gaussian integers: q divides w if and only if $w=qt$ for some Hurwitz integer t. Unfortunately, this is complicated by the fact that multiplication is not commutative, so we can't assume that $qt=tq$. So, we do the next best thing and consider the notion of *right divisibility*: if q and w are Hurwitz integers, we say that q is a *right divisor* of w if $w=tq$ for some Hurwitz integer q. In the future, whenever we say "q divides w", we will mean "q is a *right divisor* of w". As before, we write $q \mid w$ to symbolize this relationship.

Likewise, we say that a Hurwitz integer q is a *unit* if and only if q divides 1. As before, this is the case if and only if N(q)=1. Again as before, the Hurwitz integer q is called *irreducible* if it is not a unit and whenever $q=ab$ for some Hurwitz integers a and b, either a or b is a unit.

We have seen earlier in this chapter that in some sets of "integers" there is a Division Algorithm and in some there are not; we have also seen that the existence of a Division Algorithm has far-reaching consequences. So it is natural to ask whether there is a concept of divisibility with remainders for the set of Hurwtiz integers. In fact, there is, and we state that result below, with the proof omitted.

Theorem 7.11.1

If a and b are Hurwitz integers and $b \neq 0$, then there exist Hurwitz integers q and r such that $a=qb+r$, and $0 \leq \text{N}(r) < \text{N}(b)$.

The existence of a division algorithm allows us to introduce the notion of a greatest common right divisor. Specifically, we say that q is a greatest common right divisor of the two nonzero Hurwitz integers a and b if q is a common right divisor of them, and any common right divisor of them divides q. The existence of q follows from an analog of the Euclidean Algorithm. The reason this works is that whenever the Hurwitz integer t is a (right) divisor of the Hurwitz integers a and b, then t is also a (right) divisor of the remainder when the Division Algorithm is applied to a and b.

By working backward in the Division Algorithm, we also see that any gcd of a and b can be written as $qa+wb$ for some Hurwitz integers q and w. Using this fact, we can prove a version of Euclid's Lemma for the Hurwtiz integers.

Theorem 7.11.2

Let p be an ordinary integer odd prime, and suppose that a and b are Hurwtiz integers with the property that $p \mid ab$. Then $p \mid a$ or $p \mid b$.

Proof. If p does not divide a, then p and a are relatively prime, and 1 can be expressed as $1=qa+rp$ for Hurwitz integers q and r. Multiplying both sides of this equation by b gives $b=qab+rpb$. Because p is an ordinary integer, it commutes with everything, and so we have $b=qab+rbp$. It is now obvious from this expression that $p \mid b$.

Exercises

7.35 Prove that there are infinitely many quaternions q satisfying $q^2=-1$.

7.36 There are 24 units in the set of Hurwitz integers. Find them all, and prove that your list is complete.

7.37 Let us denote by L the set of quaternions with integer coefficients. The quaternions $a=1+i+j+k$ and $b=2$ are obviously elements of L. Show that if q is any element of L, then $N(a-qb) \geq 4$. Explain from this why Theorem 7.11.1 does not hold in L.

7.38 Fill in the details of the argument that the norm of a Hurwitz integer is an integer.

7.39 Prove that if q is a Hurwitz integer, then some associate of q has all-integer coefficients.

7.12 Sums of Four Squares

In this section, we use quaternions (specifically, Hurwitz integers) to prove the following theorem, referred to in Section 7.5:

Theorem 7.12.1

Any positive integer can be written as the sum of four integer squares.

We will prove this via a sequence of lemmas. Our first is the analog of Lemma 6.5.1 for four squares instead of two.

Lemma 7.12.2

If a and b are positive integers that can each be written as the sum of four squares, then ab can as well.

Proof. The proof is just like the proof of Lemma 7.5.1, this time using the fact that a and b are the norms of Hurwitz integers rather than Gaussian integers.

It is interesting to note that although we gave a quaternion-based proof of this result, it was originally proved almost a century before the quaternions were discovered by Hamilton.

It follows from the previous lemma and the Fundamental Theorem of Arithmetic that to prove that any positive integer can be written as the sum of four squares, it suffices to prove it for the numbers 1, 2 and any odd prime integer. Since the result is obviously true for the integers 1 and 2, it therefore suffices to prove Theorem 7.12.1 when n is an odd prime. To do this, we need another lemma that says that, for any prime p and any integer n, n can be written as a sum of two squares mod p. We actually don't need the result in quite this level of generality, but it is just as easy to prove it at that level.

Lemma 7.12.3

If p is any odd prime and n is any positive integer, then there are integers x and y such that $x^2 + y^2 \equiv n \pmod{p}$.

Proof. We shall work in the set $\mathbb{Z}_p$ of residue classes mod p, where for typographical convenience we denote the elements of $\mathbb{Z}_p$ as integers rather than residue classes; i.e., we write a typical element of $\mathbb{Z}_p$ as a rather than $[a]$. We need to keep in mind, however, that equality in $\mathbb{Z}_p$ amounts to congruence mod p as integers.

We will use a counting argument. We have previously seen that there are $(p-1)/2$ quadratic residues mod p. In other words, there are $(p-1)/2$ nonzero squares in $\mathbb{Z}_p$. If we add 0 to this list, we get a total of $(p-1)/2+1 = (p+1)/2$ total squares. Another way to say this is that the set $A = \{x^2 : x \in \mathbb{Z}_p\}$ has $(p+1)/2$ elements in it. It follows immediately from this that the set $B = \{n - x^2 : x \in \mathbb{Z}_p\}$ also has $(p+1)/2$ elements in it.

These observations imply that the sets A and B cannot be disjoint: if they were, then the union of these sets would contain $p+1$ elements, but that's not possible because there are only p elements in $\mathbb{Z}_p$. So, let us denote by t an element that is in both sets; t must, on the one hand, be equal to x^2 for some x in

$\mathbb{Z}_p$ and, on the other hand, be equal to $n - y^2$ for some y in $\mathbb{Z}_p$. Thus, we have $x^2 = n - y^2$ or $x^2 + y^2 = n$. Since this is an equation in $\mathbb{Z}_p$, it follows that $x^2 + y^2 \equiv n$ (mod p), as desired. This concludes the proof.

If $n = -1$, then the preceding result can be rephrased as follows: if p is any odd prime, then there exist integers x and y such that $p \mid 1 + x^2 + y^2$. This is the result that we will be using shortly. Note for future reference that $1 + x^2 + y^2 = N(q)$, where $q = 1 + xi + yj$.

We will prove that any positive integer can be written as the sum of four squares by connecting sums of four squares with Hurwitz integers. We already know that if q is a Hurwitz integer, then $N(q)$ is a nonnegative integer; it turns out that this integer is a sum of four squares. (The converse, that any sum of four squares is the norm of a Hurwitz integer, is, of course, obvious.) The argument will be simplified if we once again prove a lemma.

Lemma 7.12.4

Suppose that n is a positive integer and that $2n$ can be written as the sum of four squares. Then n can be so written.

Proof. Write $2n = a^2 + b^2 + c^2 + d^2$, where a, b, c and d are integers. Since $2n$ is even, it must be the case that a, b, c and d are either all even, all odd, or that two of them are even and two of them are odd. In any event, we may assume that (relabeling if necessary) a and b have the same parity, as do c and d. This means that the numbers $(a+b)/2$, $(a-b)/2$, $(c+d)/2$ and $(c-d)/2$ are all integers. High school algebra now shows that the sum of the squares of these four integers is n, and we are done.

We can now prove:

Theorem 7.12.5

If q is a Hurwitz integer, then the integer $N(q)$ is a sum of four squares.

Proof: We can certainly write $N(q) = a^2 + b^2 + c^2 + d^2$, where a, b, c and d are at worst half-integers. Thus $2a$, $2b$, $2c$ and $2d$ are integers, and the sum of the squares of these four integers is $4n$. Two applications of the previous lemma now establish that n is the sum of four squares.

We are now ready for the grand finale, wherein we prove that any positive integer can be written as the sum of four nonnegative squares. The good news is that the heavy lifting has all been done, and we simply have to assemble the various pieces.

Proof of Theorem 7.12.1: Recall from the remarks following Lemma 6.12.2 that it suffices to prove that any odd prime p can be expressed as the sum of four squares. From Lemma 6.12.3 (with $n = -1$), we know that there exist integers x and y such that $p \mid N(q) = q\,\overline{q}$, where $q = 1 + xi + yj$. If the ordinary integer p were also irreducible as a Hurwitz integer, this would imply $p \mid q$ or $p \mid \overline{q}$, but this is manifestly not the case. So p must be reducible, and we can write $p=ab$ where a and b are non-unit Hurwitz integers. Taking norms gives $p^2=N(a)$ $N(b)$. Since neither $N(a)$ nor $N(b)$ is equal to 1, this implies $p=N(a)$. But now we are done, because by the previous result the norm of a Hurwtiz integer is the sum of four squares.

Challenge Problems for Chapter 7

C7.1 Find all Gaussian integers α, β and γ with the property that $\alpha\beta\gamma = \alpha+\beta+\gamma=1$.

C7.2 If n is a positive integer, denote by $F(n)$ the number of Gaussian integers with norm less than n. Are there infinitely many n satisfying $F(n)=F(n+1)$?

C7.3 Prove that γ is a greatest common divisor of the nonzero Gaussian integers α, β if and only if γ is a common divisor of α, β of maximal norm.

C7.4 Use Gaussian integers to classify all integers solutions to the equation $x^2+y^2=z^3$.

C7.5 If m and n are distinct squarefree integers, prove that $\mathbb{Q}(\sqrt{m}) \neq \mathbb{Q}(\sqrt{n})$.

C7.6 Exhibit infinitely many units in $\mathbb{Z}[\sqrt{2}]$. (Hint: begin by showing that $1+\sqrt{2}$ is one.)

Appendix A: A Proof Primer

One way in which mathematics differs from all other disciplines is that in mathematics, things are *proved*—in other words, mathematics is a *deductive*, rather than *inductive*, discipline. Let us illustrate with a simple example that should be familiar to you from your high school geometry course. Consider the statement "The sum of the angles of a triangle is 180°". Suppose (contrary to fact) that you had access to a device that was capable of measuring angles with 100% precision. Suppose also that you drew 1000 triangles, all different shapes and sizes, measured the angles in each of them, and came up with an angle sum of 180° every time. Would that establish the correctness of the sentence quoted above?

The answer is *no*, for the simple reason that the angle sum of the 1001st triangle, the one that you *didn't* measure, might not be 180°. Of course it doesn't matter if we change 1000 to any other positive integer—a billion, a trillion, what have you. Since we can draw an infinite number of triangles, it is impossible to try them all; there'll always be some that we didn't measure. In order to establish the correctness of the statement, therefore, we can't simply rely on experiment; we need a *proof*.

A proof is a logically convincing argument—a series of assertions, each one with an appropriate justification, leading to the desired conclusion. We'll shortly talk about what kinds of justifications are appropriate and describe some standard kinds of proof, but before we do that we need to establish some basic vocabulary and discuss the rules of (very) elementary logic.

Many mathematical statements are what we call *conditional* statements— i.e., statements of the form "if P, then Q". This statement simply means that if we assume P, then Q must be true. The statement does not mean that Q is always true, and it says nothing at all about what happens if P is not assumed to be true. The only time a statement "if P, then Q" is false is when P is true and Q is false. So, for example, the silly-sounding statement "if Paris is the capital of Spain, then $1+1=3$" is a true statement, because the antecedent clause ("Paris is the capital of Spain") is not true. (Sentences like this are said to be *vacuously* true.) It follows from the foregoing that the negation of a conditional statement "if P, then Q" is "P and not Q".

Associated with every conditional statement "if P, then Q" is its *converse*, which is the statement "if Q, then P". (So, for example, the converse of the statement in the previous paragraph is "if $1+1=3$, then Paris is the capital of Spain".) It is important to note that the truth or falsity of a conditional statement says *nothing whatsoever* about the truth or falsity of its converse. A true conditional statement can have a converse that is true or one that is false; so can a false conditional statement. (Examples illustrating this are easy to construct, and the reader should pause now and construct some.)

Because a statement and its converse are logically independent of one another, it is wrong to assume that you can prove "if P, then Q" by assuming Q and proving P—all you would have succeeded in doing there is proving the unrelated statement "if Q, then P". This mistake, which basically amounts to assuming what you are trying to prove, is one that *many* students have made over the years, and which you should guard against.

Another statement that is related to the conditional statement "if P, then Q" is the *contrapositive* of that statement, which is "if not Q, then not P". Unlike the converse of a statement, the contrapositive *is* logically related to the original statement: it is true when, and only when, the original statement is true. So, when proving a statement, it is sometimes convenient (and perfectly acceptable) to prove the contrapositive instead. We will elaborate on this point shortly.

Closely related to conditional statements are the so-called "biconditional" statements of the form "P if and only if Q". (The phrase "if and only if" is usually abbreviated "iff".) This statement means the same thing as two statements combined: "if P, then Q" and "if Q, then P". In other words, for "P iff Q" to be true, each of P and Q must imply the other. When asked to prove an "if and only if" statement, you can give two different proofs (first prove "if P, then Q", then prove the converse) or, if you're lucky, you can write down a proof of "if P, then Q" and then note that each step in the proof is reversible; in this case, a proof can be given by a sequence of assertions, each of which is equivalent to (not just implied by) the preceding statement.

In mathematics, one frequently encounters statements that are *disjunctive* ("P or Q") or *conjunctive* ("P and Q"). For the disjunctive statement "P or Q" to be true, it suffices that either P or Q (or both) must be true. Hence, in order for this statement to be false, it must be the case that both P and Q are false. (So, for example, the *negation* of a statement "P or Q" is "Not P and not Q".) Note that disjunctive statements in mathematics are not quite like they are in ordinary English: in many situations in everyday life, the truth of a statement like "P or Q" impliedly suggests that P and Q cannot *both* be true. For example, when a mother tells a child "You can have cake or ice cream for dessert", it is implicit that a choice is being offered and that the child cannot have both. But in mathematical discourse, the truth of "P or Q" does not exclude the possibility that P and Q are both true.

A conjunctive statement "P and Q" will be true when and only when both P and Q are true; so for such a statement to be false, it suffices that either P or Q (or of course both) be false. So, for example, the statement "$1 + 1 = 3$ and Paris is the capital of France" is false, even though Paris is, indeed, the capital of France. The negation of a conjunctive statement ("P and Q") is a disjunctive one: "not P or not Q".

Two other kinds of statements must be considered here, ones that are of the form "For all..." or "There exists...". The first kind of statement is true when, and only when, it is true for all objects in the universe of discourse;

to show that such a statement is false, it therefore suffices to find one single counterexample. The statement "all prime integers are odd" is false because there is, indeed, one single example where it fails to hold: namely, the integer 2.

To prove a statement of the form "there exists..." (e.g., "there exists an even prime number"), it suffices to show that there is at least one such object. Since 2 is an even prime, merely pointing this out is sufficient to prove the statement. The fact that 2 is the *only* even prime is irrelevant to the truth of this statement; all we need to show is that there is one.

Implicit in these remarks are the facts that the negation of a "for all" statement is a "there exists" statement, and vice versa. In other words, the negation of the statement "All dog owners are happy" is NOT "All dog owners are unhappy"; it is, instead, "There exists an unhappy dog owner".

Sometimes, an existence theorem can be proved without explicitly giving an example of the desired object. This is called a non-constructive proof. Here is an amusing example of such a proof. We want to prove that there exist two irrational numbers α and β with the property that α^β is rational. (An irrational number is one that cannot be expressed as a quotient of integers; it is a fact, one that is proved in the text and will be assumed here, that $\sqrt{2}$ is irrational.) For the proof, consider the number $\gamma = \sqrt{2}^{\sqrt{2}}$. If γ is rational, take $\alpha = \beta = \sqrt{2}$, and we are done. If γ is not rational, take α to be γ, and β to be $\sqrt{2}$. Then $\alpha^\beta = \sqrt{2}^{\sqrt{2}\sqrt{2}} = (\sqrt{2})^2 = 2$, which is rational. So either way we have found two irrational numbers α and β with the property that α^β is rational. (It actually turns out that γ is not rational, but that fact is quite hard to prove. The point is that we don't need to know whether it is or not for this proof to work.)

The previous proof illustrates another technique that is often used in proofs—consideration of cases. Occasionally, while working one's way through an argument, one encounters a situation that can occur in multiple ways. In such a situation, it may be useful to just consider each possible way the situation can occur and show the result is true in each case.

We now turn to the mechanics of proof in general. As stated earlier, a proof consists of a string of assertions, each appropriately justified, leading to a desired conclusion. (In high school geometry you probably did "two column proofs" where each assertion really was a separate line in a column, with the justification in the second column. Mathematicians write proofs in prose, but it may help you to first write a two-column proof and then work on putting the lines together into prose.) There are six permissible justifications for a line in a proof: definition, assumption, axiom, previously proved theorem, previous line in a proof, or principle of logic. Of these, the one that probably requires the most explanation is "axiom".

Modern mathematics is often done axiomatically: i.e., certain principles are taken as "given" (they are, so to speak, the "rules of the game") and deductions are made from them. You may have encountered this in your geometry classes: the statement "given any two points, there is a unique line containing them" is often taken as an axiom of Euclidean geometry. It isn't something that we attempt to justify rigorously; we simply *assume* it to be true. (Words like "point" and "line" are generally taken as undefined terms; since there are only a finite number of words in the English language, it is impossible to define everything; if you tried, you would eventually wind up in a circular situation.)

In this number theory book, we have not attempted to rigorously define the set of integers by specifying axioms for them. We simply assume the reader is familiar with them, and we assume as known all the familiar facts from arithmetic that the reader has used for years. However, it is worth noting that some of these facts can be taken as axioms and others can be proved as consequences of these axioms. Appendix B summarizes some axioms for the integers and also specifies some of the results that can be deduced, as theorems, from these axioms.

We now turn to a survey of some basic methods of proof. First is the direct method, where, to prove "if P, then Q" we simply assume P and proceed, step by step and using the six basic justifications specified above, to prove Q. We illustrate with an example. In Chapter 1 of this text we define, for two integers m and n, the relation "n divides m" (denoted $n \mid m$) to simply mean $m = nx$ for some integer x; this is just a precise way of saying "n goes evenly into m". The following theorem, summarizing basic facts about divisibility, is stated in Chapter 1; we prove it here. The proofs are quite easy but do illustrate the method of a direct proof.

Theorem A.1

If m, n and r are integers, then the following are true:

(a) $n \mid n$

(b) $1 \mid m$

(c) if $n \mid m$ and $m \mid r$ then $n \mid r$

(d) if $n \mid m$ and $n \mid r$ then $n \mid m + r$ and $n \mid m - r$

Proof

(a) We know that $n = n1$, and 1 is an integer. Therefore, by definition of divisibility, $n \mid n$.

(b) We know that $m = 1m$. Therefore, by definition of divisibility, $1 \mid m$.

(c) By definition, we know that there exist integers s and t such that $r = sm$ and $m = nt$. It therefore follows that $r = s(nt) = n(st)$. Since st is an integer, it follows by definition that $n \mid r$.

(d) By assumption, there exist integers s and t such that $r=sn$ and $m=nt$. So $m+r=sn+nt=n(s+t)$. Since $s+t$ is an integer, it follows by definition that $n \mid m+r$. The same argument shows that $n \mid m-r$.

Another method of proof is *proof by contradiction*. The idea here is that, to prove "if P, then Q", we assume P but also assume that Q is false, and, from these two assumptions, derive some kind of contradiction. Since the assumption that Q is false leads to a contradiction, it must therefore be the case that Q is true, which is what we wanted to prove.

Students tend to overuse proof by contradiction. Some students have even been known to assume the negation of Q, then prove Q directly, and then argue that they have found a contradiction. Of course this is wasted work: if you can prove Q directly, you don't need a proof by contradiction!

As a simple illustration of proof by contradiction, we prove a few other properties of divisibility that will be used throughout the book.

Theorem A.2

If m, n and r are integers, then the following are true:

(a) if $n \mid 1$ then $n=\pm 1$

(b) if $n \mid m$ and $m \mid n$ then $n=\pm m$

Proof

(a) We are told that there exists an integer x such that $1=nx$. It is intuitively obvious that this forces x to be 1 or -1, but let's give a more careful proof, using the fact that 1 is the smallest positive integer. (The fact that 1 is, indeed, the smallest positive integer is something that you can assume for the moment, but we will give a precise proof immediately after the conclusion of this one.) Since $1=nx$, it is clear that x is nonzero, so is either positive or negative. If x is positive and not equal to 1, it is greater than 1, but then (since n must be positive as well) we have $nx>x>1$, contradicting our assumption that $1=nx$. Finally, suppose x is negative. Then $1=nx=(-n)(-x)$, where now $-x$ is positive. By what we have just done, this forces $-x=1$, from which we conclude $x=-1$, as desired.

(b) Try this yourself.

As another example of the method of proof by contradiction, we prove a few consequences of the Well-Ordering Principle, namely (see Appendix B) that any nonempty set of positive integers has a smallest element. We first prove a result that seems almost insultingly obvious: that there is no integer between 0 and 1, or, to rephrase things, that 1 is the smallest positive integer. Although obvious sounding, this result is actually used in other proofs (in fact, we just used it above, and will again use it, almost immediately, to prove the Principle of Mathematical Induction) and, if you're going to do things very precisely, requires proof. The proof is actually quite simple and provides a good illustration of how to use the Well-Ordering Principle.

Theorem A.3

There is no positive integer that is less than 1.

Proof

Suppose to the contrary that a positive integer less than 1 existed. Then the set S of all positive integers less than 1 is nonempty, and hence, by the Well-Ordering Principle, has a smallest element; call it x. Multiply the inequality $0 < x < 1$ by x; since we are multiplying by a positive integer, the inequality is preserved and we get $0 < x^2 < x < 1$. It follows from this that x^2 is a positive integer that is less than 1 but also less than x, which contradicts our definition of x.

We next prove the Principle of Mathematical Induction (see Section 1.1 of the text) as a consequence of the Well-Ordering Principle. For convenience, we restate the Principle of Mathematical Induction.

Theorem A.4

(Principle of Mathematical Induction). Suppose that

- S is a subset of the set of positive integers,
- $1 \in S$, and
- $n + 1 \in S$ whenever $n \in S$. Then S consists of all positive integers.

Proof

Assume, hoping for a contradiction, that there is a positive integer that is *not* in S. Then, by the Well-Ordering Principle (applied to the nonempty set of all such integers), there must be a *smallest* positive integer not in S; call it k. Note that $k \neq 1$ (because $1 \in S$), so $k - 1$ is a positive integer. (Note that we are using the previously proved result here!) It is also in S (since it is smaller than k.)

However, by assumption, since $k-1 \in S$, it must be the case that $k=(k-1)+1 \in S$, a contradiction. This contradiction yields the desired result.

To see other examples of the power of the method of proof by contradiction, refer to Euclid's proof, reproduced in the text, that there are infinitely many primes, and also the proof in the text that $\sqrt{2}$ is irrational.

Closely related to the method of proof by contradiction is the method of *proving the contrapositive*. Recall that the contrapositive of a statement is logically equivalent to the original statement, so to prove "if P, then Q" it suffices to prove "if not Q, then not P". In other words, it suffices to prove that if Q is false, then P is false as well. This actually amounts to a proof by contradiction, since if we can deduce the negation of P and are assuming P to be true, then we have a contradiction. However, it is a very special kind of proof by contradiction and, unlike the method of proof by contradiction, applies only to conditional statements.

As a simple illustration of this method, we prove an easy result about even and odd integers. For purposes of this proof, we will assume as known that any integer is either even or odd, that the even integers are precisely those that can be written as $2n$ for some integer n, and that the odd integers are precisely those that can be written as $2m+1$ for some integer m.

Theorem A.5

If a is an integer and a^2 is even, then a is even.

Proof

It suffices to prove the contrapositive—i.e., that if a is odd, then a^2 is odd. Suppose, therefore, that a is odd. Then we can write $a=2m+1$ for some integer m. But then $a^2=(2m+1)^2=4m^2+4m+1=2(2m^2+2m)+1$. Since $2m^2+2m$ is an integer, call it n, we have written a^2 as $2n+1$ for some integer n, which means that a^2 is odd, as desired.

Exercises

A1. Is the statement "Paris is the capital of France and New York is the capital of Spain" true or false? What is the negation of this statement?

A2. Write down the negation of the statement "If it is raining, I will go to the movies".

A3. Prove part (b) to Theorem A.2 above.

A4. Write down four true statements, two of which have false converses and two of which have true converses. The statements you choose can be "mathematical" or "nonmathematical", as you choose.

A5. For purposes of this problem, assume that any integer can be written in the form $2m$ or $2n+1$ for some integer m or n. Integers of the first kind are, of course, called even; integers of the second kind are called odd. Use properties of divisibility to prove that no integer can be both even and odd.

A6. (See previous problem.) Prove that the sum of two even, or two odd, integers is even. Prove that the sum of an even integer and an odd integer is odd.

Appendix B: Axioms for the Integers

Because the reader has presumably been dealing with the set of integers for years now, he or she is no doubt familiar with some of their very basic properties—for example, that the product of two nonzero integers is nonzero. In this book, we will simply assume familiarity with these properties and use them freely. However, it is worthwhile to note that the set of integers can be characterized by axioms, or assumptions, that, if taken for granted, can be used to *prove* all the other properties of the integers that we will need. For the benefit of those who prefer a somewhat more formal approach to the integers, and to give some practice in the construction of simple proofs, we briefly indicate in this Appendix how an axiomatic approach can be carried out. Our axioms are divided into three groups: axioms of Arithmetic, Order and a Well-Ordering Principle. We assume the existence of the set $\mathbb{Z} = \{...-2, -1, 0, 1, 2, ...\}$ on which are defined two operations of addition and multiplication.

Arithmetic Axioms: if m, n and r denote arbitrary integers, then

1. $m + n = n + m$ (commutative law for addition)
2. $(m + n) + r = m + (n + r)$ (associative law for addition)
3. the integer 0 satisfies $m + 0 = m$ (existence of additive identity)
4. the integer 1 satisfies $1m = m$ (existence of multiplicative identity)
5. for every m, the integer $-m$ satisfies $m + (-m) = 0$ (existence of additive inverse)
6. $mn = nm$ (commutative law for multiplication)
7. $(mn)r = m(nr)$ (associative law for multiplication)
8. $m(n + r) = mn + mr$ (distributive law)
   (On the basis of these axioms, we can define subtraction as follows: $a - b = a + (-b)$.)

Order Axioms: there exists a nonempty subset $\mathbb{P}$ of the set of integers, called the set of positive integers, with the following properties:

9. if m is an arbitrary integer, then exactly one of the following holds: $m \in \mathbb{P}, -m \in \mathbb{P}, m = 0$ (trichotomy law)
10. if $m, n \in \mathbb{P}$, then $mn \in \mathbb{P}$ and $m + n \in \mathbb{P}$ (closure)
    (On the basis of these axioms, we can define a *negative* number to be an integer m with the property that $-m$ is positive. We can also define an order relation $<$ as follows: $a < b$ means that $b - a \in \mathbb{P}$. In an analogous way, we can define the relations $>$, $\leq$ and $\geq$.)

Well-Ordering Principle:

  11. Any nonempty set S of positive integers contains a smallest element, i.e., an element x with the property that $x \leq y$ for all $y \in$ S.

On the basis of these axioms, one can prove all the standard facts about integer arithmetic that were learned in grade school and routinely used since then. These include (for integers m, n and r):

- If $m + r = n + r$, then $m = n$
- $m0 = 0$
- $-(-m) = m$
- $(-m)(n) = -mn$
- $(-m)(-n) = mn$
- if $mn = 0$ then either $m = 0$ or $n = 0$
- if $mr = nr$ and r is nonzero, then $m = n$

The reader may wonder why it is even necessary to prove "obvious" facts like these. This is the nature of mathematical reasoning: when proving things from axioms, we cannot take anything for granted. If one is going to develop mathematics rigorously, then careful definitions and careful proofs (even of things that seem obvious) cannot be avoided. So, for the sake of completeness, we will prove some of the facts above and leave the others as exercises.

We start by proving the first property, which can be summarized by the phrase "additive cancellation". Suppose $m + r = n + r$. Then add $-r$ to both sides of this equation, getting $(m + r) + (-r) = (n + r) + (-r)$, which by the associative law reduces to $m + (r + (-r)) = n + (r + (-r))$, which by axiom 5 leads to $m + 0 = n + 0$, or (by axiom 3) $m = n$, as desired.

With this established, we can easily prove the second property above. We know that $0 = 0 + 0$ by axiom 3, so we have the following chain of equalities: $0 + m0 = m0 = m(0 + 0) = m0 + m0$. It follows from this, and the previous result, that $m0 = 0$.

For the third property, first consider $m + (-m)$. By axiom 5, this is 0. On the other hand, axiom 5 also tells us that when we add $-(-m)$ to $-m$, we get 0. So we have

$-(-m) + (-m) = m + (-m)$, and by additive cancellation, it follows that $-(-m) = m$.

We leave it to the reader to prove the fourth and fifth properties above. To prove the sixth property above, we use the order axioms as well as the arithmetic ones. If $mn = 0$ and neither m nor n are zero, then there are three possibilities: both m and n are positive, both are negative, or one is positive and one is negative. If m and n are positive, then by axiom 10 the product mn is also positive, and hence, by axiom 9, cannot be 0. If m and n are both negative, then $-m$ and $-n$ are positive, and once again $mn = (-m)(-n)$ is positive, and hence can't be 0. We leave to the reader the task of disposing of the one remaining case and also proving the seventh and last bulleted property above.

One thing that might be noted from the list of results above is that the familiar fact that "the product of two negative numbers is positive", a fact that students learning arithmetic for the first time sometimes wonder about. This fact, we now see, actually follows logically from the other axioms. A number of other "obvious" facts about arithmetic follow from these definitions, but aside from a few that are listed in the exercises (e.g., $-0 = 0$), we will not make the effort to list and prove all of them; now that we have given a set of axioms and seen how they can be used, we will simply take all the familiar basic principles of arithmetic as given and use them without explicit proof.

Note also that nothing is said in these axioms about division. There's a reason for that, of course: there is no operation of division defined on the integers because the quotient of two integers may very well not be an integer. For example, 1 divided by 2 is ½, which is certainly not an integer. However, we will see in the text that given two integers, we can divide one by the other, obtaining a quotient and remainder. This is another "intuitively obvious" result, and the reason it is not listed as an axiom is that it can be deduced, as a theorem, from the other axioms. Here, however, the proof is not trivial, but it is instructive, so it is proved in Chapter 1. Another nontrivial but very useful result that can be deduced from the axioms is the Principle of Mathematical Induction, which is also discussed in Chapter 1.

Exercises

B1. Prove the fourth, fifth and seventh bulleted properties of the integers.

B2. Prove that if m is a nonzero integer then $m^2 > 0$.

B3. Prove that if $a < b$ and $b < c$ then $a < c$.

B4. Explain from the axioms why $-0 = 0$.

Appendix C: Basic Algebraic Terminology

Although it is not technically necessary to know abstract algebra in order to understand the basic ideas of elementary number theory, it turns out that a number of these number-theoretic ideas are best understood in an algebraic context. So, in this Appendix, we introduce some of the basic terminologies of abstract algebra and give some examples of these ideas. Proofs of the results that are stated here can be found in any undergraduate abstract algebra textbook.

First, we introduce the notion of a *group*. A *group* is an ordered pair (G, *) where G is a set and * is a binary operation on G (i.e., a function that associates to any ordered pair (a, b) of elements of G an element $a*b$ in G) that satisfies (for all elements a, b and c in G) the following properties:

- $(a*b)*c = a*(b*c)$ (associativity)
- there exists an element e in G with the property that for all $a \in$ G, $a*e = e*a = a$ (identity element)
- for every $a \in$ G, there exists an element, denoted a^{-1}, such that $a* a^{-1} = a^{-1} *a = e$ (inverse element)

When the binary operation * is understood, we will denote the group just by identifying the set and speak of "the group G". In addition, it is customary to suppress the * notation and denote the binary operation by ordinary juxtaposition of letters. In other words, we write ab instead of the more cumbersome $a*b$. It should be kept in mind, however, that ab does not necessarily symbolize the product of a and b under any kind of multiplication, but instead the product under an abstract operation.

One other point should be emphasized: it is implicit in the definition of "binary operation" that the set G is closed under the binary operation *: in other words, if a and b are elements of G, then $a*b$ is also an element of G. Thus, for example, the set of positive integers is not a group under subtraction, because the set is not closed under this operation: 3 and 5 are in the set, but $3-5 = 2$ is not.

A fairly trivial consequence of the defining conditions of a group is that in any group G, cancellation holds: if $ab = ac$, then $b = c$. To see this, simply "multiply" both sides of the given equation by a^{-1} on the left and use the associative law.

Some more definitions: If G is a finite set, with, say, n elements, then we say G has *order n*; if G is an infinite set, then we say G has *infinite order*. If the binary operation is commutative, i.e., $ab = ba$ for all a, $b \in$ G, then we say that G is an *abelian* group. (This is named for the Norwegian mathematician Neils Hendrik Abel.)

We can define exponentiation of group elements as follows: if G is a group, $a \in G$, and n is a positive integer, then a^n simply means the "product" of a with itself n times. If n is negative, then $a^n = (a^{-n})^{-1}$. Finally, we define $a^0 = e$, the identity of G. With these definitions, it can be shown, via a fairly tedious inductive argument, that all the usual "rules of exponents" hold. For example, $a^n a^m = a^{n+m}$ for all integers n and m.

We list here some examples of groups. The first two are particularly relevant to the study of elementary number theory.

- Let $G = \mathbb{Z}$, the set of integers, and define $a * b = a + b$. In other words, the binary operation is ordinary addition. Then G is an abelian group with identity element 0; the inverse of an integer a is $-a$. Note that the same set G, with respect to the operation of multiplication, is *not* a group (why?).

- If $G = \mathbb{Z}_n$, the set of congruence classes modulo n (see Chapter 2 for the definition), then G is also a group under congruence class addition. G is not a group under multiplication, however. (The congruence class [0] has no inverse.) If we consider the set of *nonzero* congruence classes modulo n, then this is a group under congruence class multiplication if and only if n is a prime. The reader should verify these facts, and note the connection with Euclid's Lemma.

- Both groups specified above are abelian. For a non-abelian example, let G be the set of $n \times n$ nonsingular matrices with real entries. This is a group under matrix multiplication, but, as is easily shown, it is not abelian.

If H is a subset of a group G that, with the same operation that makes G a group, is itself a group, then we say H is a *subgroup* of G. Example: the set of even integers is a subgroup of the set $\mathbb{Z}$ of all integers, as are $\mathbb{Z}$ itself and, on the other extreme, the one-element set $\{0\}$. The set of odd integers is not a subgroup of $\mathbb{Z}$; however, for several reasons, the first being that addition is not even a binary operation on this set: the sum of two odd integers is even, not odd, and hence is not in the set. Another reason why this subset is not a subgroup is the fact that there is no identity (0 is not odd).

The first significant result that one learns in a course on group theory is *Lagrange's Theorem*, which states that if G is a finite group of order n and H is a subgroup of order m, then m divides n. The converse is not true: if G is a group of order n and m is a positive integer that divides n, it is not necessarily the case that G contains a subgroup of order m. However, constructing a counterexample is not a trivial undertaking and doing so now would take us too far afield.

Lagrange's Theorem does, however, have a number of corollaries that are of interest in the study of elementary number theory. To discuss these, we need some more definitions. Suppose that G is a group and that a is an element of G.

Consider the set of nonnegative powers of a: $e, a, a^2, \ldots$. One of two things must be the case: either these powers are all distinct or there is some repetition among them, say $a^m = a^n$ with $m > n$. If the latter condition holds (which it must if G is finite), then by cancellation $a^{m-n} = e$, and so by the Well-Ordering Principle there is a smallest positive integer d such that $a^d = e$. This smallest positive integer d is called the *order* of a. If the powers of a are all distinct, then we say that a has *infinite order*.

Now, suppose a has order d. Then it is not hard to see that the set $\{e, a, a^2, \ldots, a^{d-1}\}$ is a subgroup of G of order d; let us denote this set $< a >$. Note that all higher powers of a are automatically in $< a >$: since $a^d = e$ by assumption, we "loop around" when considering a^d, $a^{d+1} = a^d a = a$, etc. Observe also that this set is the smallest possible subgroup of G containing a; we call it the *subgroup of G generated by a*. Thus, if an element of a group has finite order, this order is also the order of the subgroup generated by that element. It follows that if G is a finite group of order n, then (by Lagrange's Theorem), $d \mid n$. This observation, in turn, allows us to deduce another: since $n = dk$ for some integer k, it follows that $a^n = a^{dk} = (a^d)^k = e^k = e$. Thus, in a group of order n, *if we take any element and raise it to the nth power, we get the identity.*

We next consider a different kind of algebraic system, one with *two* binary operations defined on a set R. These operations are called addition (denoted +) and multiplication (denoted by juxtaposition). We say that R, with respect to these operations, is a *ring* if, for all a, b and c in R:

- The set R is an abelian group with respect to addition (with the identity denoted 0)
- The distributive laws hold: $a(b+c) = ab + ac$ and $(b+c)a = ba + ca$
- Multiplication is associative: $(ab)c = a(bc)$
- There is a multiplicative identity, i.e., an element 1 in R such that $1a = a1 = a$.

A few remarks: First, not all authors require the last condition (multiplicative identity) as part of the definition of a ring and use the term *ring with identity* to denote rings that happen to have a multiplicative identity. However, it is becoming more and more common to require a ring to have an identity, and since the rings that we will encounter all do have an identity, we will require this condition as part of the definition.

Second, note that we have not required multiplication to be commutative. In other words, we do not require that $ab = ba$ for all elements a and b in R. A *commutative ring* is one in which this requirement does hold.

Here are some examples that will be particularly relevant for us. The set $\mathbb{Z}$ of integers is a commutative ring with respect to the "usual" operations of addition and multiplication, as is the set $\mathbb{Z}_n$ of congruence classes modulo some positive integer n. Likewise, the sets $\mathbb{Q}$ and $\mathbb{R}$ of rational and real numbers, respectively, are commutative rings. The set of even integers is not a ring under our definition because it lacks a multiplicative identity (1 is not even). The set of $n \times n$ matrices with real entries is a ring under the usual operations of matrix addition and matrix multiplication but is not a commutative ring, because it is easy (for $n > 1$) to find $n \times n$ matrices A and B for which $AB \neq BA$.

If R is a ring (with identity 1), an element $a \in R$ is called a *unit* if there is an element $b \in R$ with the property that $ab = ba = 1$, the multiplicative identity of R. In other words, the units of a ring are those elements that have multiplicative inverses. In ring $\mathbb{Z}_6$, for example, the units are 1 and 5. In general, the units of $\mathbb{Z}_n$ are those elements that are relatively prime (see Chapter 1 for the definition) to n.

A commutative ring in which every nonzero element is a unit is called a *field*. It follows from the previous observation that $\mathbb{Z}_n$ is a field if and only if n is a prime. The ring $\mathbb{Z}$ of integers is not a field because, for example, 2 is not a unit; this is because ½ is not an integer. The ring of Gaussian integers $\mathbb{Z}[i]$ (see Chapter 7) is, likewise, not a field: indeed, as established in Chapter 7, the only units in that ring are 1, −1, i and −i. The rings $\mathbb{Q}$ and $\mathbb{R}$ are fields, however.

If F is any field and a and b are nonzero elements of F, then ab must also be nonzero: if the contrary were true and $ab = 0$, multiplying both sides of this equation by the multiplicative inverse of a would give $b = 0$, a contradiction. Hence, the set F^* of nonzero elements of F is closed under multiplication. (Algebraists say that F has no nonzero zero divisors.) From here it is easy to see that F^* is in fact an abelian group under multiplication and that, therefore, F^* has the cancellation property. There are non-fields that satisfy this, however; $\mathbb{Z}$ and $\mathbb{Z}[i]$ are two examples. These are examples of algebraic structures called *integral domains*: commutative rings that have no nonzero zero divisors, or, equivalently, satisfy the cancellation property. So, while any field is an integral domain, it is not necessarily the case that any integral domain is a field (although it is a standard exercise in abstract algebra textbooks to establish that any finite integral domain is a field).

Bibliography

[AC] A. Adler and J.E. Coury, *Theory of Numbers: A Text and Source Book of Problems*, Jones and Bartlett Publishers, Burlington, MA, 1995.

[Cam] D. Campbell, *An Open Door to Number Theory*, MAA Press, New Denver, 2018.

[Con] K. Conrad, *The Gaussian Integers*, https://kconrad.math.uconn.edu/blurbs/ugradnumthy/Zinotes.pdf.

[Jar] F. Jarvis, *Algebraic Number Theory*, Springer-Verlag, Heidelberg/Belin, Germany, 2014.

[Kim] S. Kim, "An elementary proof of the quadratic reciprocity law," *American Mathematical Monthly*, 111, 1 (2004), 48–50.

[KW] J. Kraft and L. Washington, *An Introduction to Number Theory with Cryptography*, 2nd edition, CRC Press, Boca Raton, FL, 2018.

[NZM] I. Niven, H. Zuckerman and H. Montgomery, An Introduction to the Theory of Numbers, 5th edition, Wiley, New York, 1991.

[R-S] S. Rubinstein-Salzedo, *Cryptography*, Springer-Verlag, Heidelberg/Belin, Germany, 2018.

[Ros] K. H. Rosen, *Elementary Number Theory*, 6th edition, Pearson, Upper Saddle River, NJ, 2010.

[Sil] J. Silverman, *A Friendly Introduction to Number Theory*, 4th edition, Pearson, Upper Saddle River, NJ, 2011.

[St] J. Stillwell, *Elements of Number Theory*, Springer-Verlag, Heidelberg/Belin, Germany, 2003.

Index

Printed in the United States
by Baker & Taylor Publisher Services

Printed in the United States
by Baker & Taylor Publisher Services